Wandering Continents and
Spreading Sea Floors
on an Expanding Earth

Wandering Continents and Spreading Sea Floors on an Expanding Earth

Lester C. King
University of Natal, Durban, South Africa

A Wiley-Interscience Publication

JOHN WILEY & SONS
Chichester · NewYork · Brisbane · Toronto · Singapore

Library of Congress Cataloging in Publication Data:

King, Lester Charles.
  Wandering continents and spreading sea-floors on an expanding earth.
  'A Wiley–Interscience publication.'
  Bibliography: p.
  Includes index.
  1. Continental drift.   2. Sea-floor spreading.
I. Title.
QE511.5.K53   1983   551.1′36   83-1345

ISBN 0 471 90156 3

British Library Cataloguing in Publication Data:

King, Lester C.
  Wandering continents and spreading sea-floors on an expanding earth.
  1. Continental drift.
  I. Title.
  551.1′3   QE511.5

ISBN 0 471 90156 3

Phototypeset by Input Typesetting Ltd, London SW19 8DR
and printed by the Pitman Press, Bath, Avon

Contents

Foreword

A geophysical Credo

The conference on the 'Implications of Continental Drift to the Earth Sciences' at Newcastle in 1972 was drawing to a close. For a week, participants had discussed new hypotheses on continental drift, rift valleys, sea-floor spreading, plate tectonics, reversals of geomagnetic polarity and related topics, and now they were listening to the final paper, by Dr R. S. Dietz, on 'Continents adrift: new orthodoxy or persuasive joker'. Ably illustrated by J. C. Holden, the paper was a glorious spoof, and like all good spoofs it left the hearers uncertain as to how much was good clean fun and how much was underlying doubt upon those matters which the Conference had debated. Dietz himself had been a major contributor in fields of the 'New Tectonics'; on maturer thought, and with further data, how much of the bright new hypotheses had he come to doubt? He ended with a quotation from Scharnberger and Kern (*Geotimes*, 1972):

THE GEOTECTONICS CREED
I believe in Plate Tectonics Almighty, Unifier of the Earth Sciences, and explanation of all things geological and geophysical; and in our Xavier le Pichon, revealer of relative motion, deduced from spreading rates about all ridges; Hypothesis of Hypothesis, Theory of Theory, Very Fact of Very Fact; deduced not assumed; Continents being of one unit with the Oceans, from which all plates spread; Which, when they encounter another plate and are subducted, go down in Benioff Zones, and are resorbed into the Asthenosphere, and are made Mantle; and cause earthquake foci also under Island Arcs; They soften and can flow; and at the Ridges Magma rises again according to Vine and Matthews; and ascends into the Crust, and maketh symmetrical magnetic anomalies; and the sea floor shall spread again, with continents, to make both mountains and faults, Whose evolution shall have no end.
  And I believe in Continental Drift, the Controller of the evolution of Life, Which proceedeth from Plate Tectonics and Sea-Floor Spreading; Which with Plate Tectonics and Sea-Floor Spreading together is worshipped and glorified; Which

was spake of by Wegener; And I believe in one Seismic and Volcanistic pattern; I acknowledge one Cause for the deformation of rocks; and I patiently look for the eruption of new Ridges and the subduction of the Plates to come. Amen.

In ensuing chapters we shall review some of the beliefs covered by the 'New Tectonics'; and the evidence relating to them.

Preface

To date I have published three books all dealing with the landforms and history of the Earth's surface. *South African Scenery* (1942) was the work of a young academic concerned to understand and to teach of the landforms in the country where he had made his home.

The Morphology of the Earth (1962) reviewed the landscapes of all the continents. It sought and expounded a common geomorphic history embracing short, active episodes of tectonic elevation or depression, alternating with prolonged intermissions of denudation synchronously in all the continents. To gather his data, the author worked in more than fifty countries, in all the seven continents, from the Arctic Circle to the South Pole. '*The Morphology*' occupied twelve years in the writing.

Third came *The Natal Monocline: Explaining the Origin and Scenery of Natal, South Africa* (1972), a miniature which showed that the province wherein I had dwelt for nearly forty years exhibited all the main chapters (or cycles) of landscape evolution from the Jurassic period to the present day. Natal is, indeed, a rare geomorphic gem, and the topographic evolution of this province may well serve as a standard for the world.

And now, as Emeritus, I take up pen once more to re-scan the face of the globe, that we may draw from its Earthscapes, both above and below sea level, information and conclusions regarding the later tectonics of our planet. For this survey, much information comes from beneath the sea, so may I here express due admiration for, and indebtedness to, the host of younger scientists who have provided such a wealth of data from research ships at sea.

Lester King
Geology Dept.
University of Natal,
King George V Avenue,
Durban 4001

Chapter 1

Introducing Gondwana and Laurasia

The beginning

The planet Earth came into existence through a cosmic event about 4500 million years ago.

The oldest stratigraphic systems, recording the existence of solid crust, appeared 1000 million years later. Their sedimentary rocks indicate environments in large water bodies, perhaps primitive oceans. At that time the crust was thin and nucleated (Fig. 1); so that the sediments were extensively intruded by masses of granite and by greenstone belts of basic and ultrabasic magmas discharged from the underlying mantle zone which forms the body of the Earth. Sedimentary rocks formed even about this time (3550 Ma)* prove by contained fossils and amino acids the existence already of living organisms upon the Earth.

Amid much crustal disturbance, and with time, the crustal nuclei thickened and clustered until by approximately 2500 million years ago the present volumes and areas of the continents had been welded together. These amounts of material have subsequently been conserved through Earth history although their distribution about the globe has changed from time to time. Their vicissitudes form the subject of geological study.

Gondwanaland and Laurasia

By the late Precambrian (1000 million years ago or less), all the available continental material had aggregated into two ovoid supercontinents called Gondwanaland and Laurasia, situated in the southern and northern hemispheres respectively (Fig. 2). The reality of Gondwanaland in the geologic past was first demonstrated by the master geologist Alex L. du Toit (1937) who compared in the field the Palaeozoic and Mesozoic rocks of South America with those of southern Africa. He found that formation after forma-

---

* 1 Ma = 1 million years.

2

Figure 1   Ancient continental crust showing growth by accretion of primitive domed
nuclei, Rhodesia (after A. M. MacGregor)

tion, in ordered sequence for thousands of metres thickness, and covering
thousands upon thousands of square kilometres of territory, were almost
exactly the same. Moreover, they were disposed in similar, simple structures,
and could by no possible means have originated with the continents in their
present positions 6000 km apart and separated by the South Atlantic Ocean.
During the Palaeozoic and early Mesozoic eras the two continents must have
lain nearly side by side and their comparable formations (similar even in
their igneous intrusion and fossil content (including quadrupeds)) must have
accumulated in contiguous and continuous continental basins (Figs 2, 3).

Conversely, du Toit's demonstration that the Palaeozoic continents were
not positioned as they are now was the first detailed geological proof of
the reality of continental drift during the interim (Chapter 5). He showed,
moreover, that where there are differences of formation (e.g., phasal variations in sedimentary formations) these find logical and proper explanation
within the Gondwanaland concept.

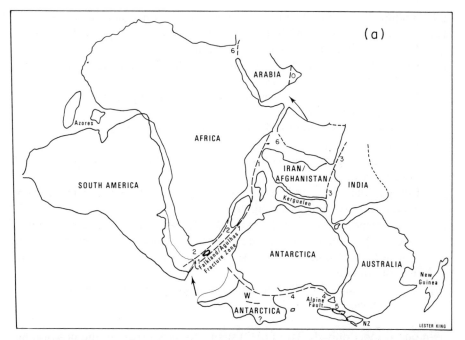

Figure 2(a)   Reassembly of Gondwanaland from stratigraphic structural data. A broad continental shelf fringes the southern end of South America. Madagascar parallels southeast Africa between Beira and Durban, with the Mascarene submarine plateau similarly disposed above. Several well-known fracture zones are clearly related to the break-up, *viz.*, trans-Antarctic fracture zone, Falkland–Agulhas and Prince Edward fracture zones, and the Baluchistan fracture zone

In 1937 du Toit reviewed the correlations for all the southern continents and showed that, for Palaeozoic/Mesozoic time, they could be neatly reassembled into one supercontinental Gondwanaland with coherent formations and structures throughout (Fig. 2). The scope, magnitude and detail of du Toit's criteria for the transoceanic comparison of segments of Gondwanaland are seldom fully appreciated. They covered almost every phase of geology visible to a field worker and in sum total they provide the original and conclusive proof (on the basis of demonstration by field facts) of the reality of post-Triassic continental drift. They include:

(a) General similarity of coastlines.
(b) Fracture patterns as fault-line scarps and monoclines.
(c) Abruptly terminated plateaus or elevated erosional plains with disturbed or reversed drainages.
(d) Submarine features such as the mid-Atlantic and mid-Indian Ocean ridges.
(e) Equivalent formations on opposite coasts with due regard to their mode of origin and subsequent history.

4

(f)   Similar variation in formations when traced along opposite shores.
(g)   Contrasted phasal variation when traced away from opposed shores.
(h)   Unconformities on comparable horizons.
(i)   Comparable geosynclinal troughs in each mass having more or less similar trends and histories.
(j)   Comparable fold systems passing out to sea at opposed shores.
(k)   Crossings of fold or fault systems of specific ages.
(l)   Synchronous intrusion of batholiths in equivalent fold systems.
(m)   Plateau basalts and associated dyke swarms.
(n)   Petrographic provinces with similar eruptive suites belonging to several different ages.
(o)   Comparable zones and periods of ore-genesis, especially such as may contain rare or distinctive minerals.
(p)   Strata denoting a special environment, particularly extreme climatic types such as tillite, varved shale, laterite, evaporite, aeolian sandstone, coal and coral limestone. These must be fitted into a rational scheme of past climates.
(q)   Terrestrial faunal and floral palaeontologic provinces with identical or allied species.

In one respect only du Toit (1937) seems to have erred: in the position he ascribed to Madagascar; he placed it well to the north in the latitude of Kenya and Tanzania, largely on the basis of a similarity of Precambrian strikes and an absence of the Jurassic basalts so typical of southern Africa.

A much better reassembly was made by Flores* (1970), who placed it in the range between Beira and Durban.

Both Kent (1972) and Tarling (1972) accepted this, with Creer remarking that the northern position placed 'the Madagascan pole 17, lies in Patagonia, suggesting that the position allotted to Madagascar in the Smith and Hallam (1970) reconstruction of Gondwanaland (which followed due Toit) is not correct.'

On the present continents there are several occurrences where the sedimentary materials of the late Palaeozoic–early Mesozoic Systems were shed on to their sites from beyond the present boundaries of those continents. This is so — with directions of derivation — in southern Brazil (from the southeast), in Zaire (from the northwest), in the Cape Province (from the south), in Natal (from the northeast), in Orissa, India (from the southeast). The materials are distinctively of continental type derived from granite–gneiss terrains and cannot be attributed to minor topographies such as volcanic island arcs. When the reconstruction is made the problems of extracontinental derivation find immediate solution within the Gondwanaland supercontinent itself.

Through the Palaeozoic era, the two subequal, ovoid supercontinents

* Chief Geologist, Moçambique Gulf Oil.

contained between them all the known land area of the globe. Because they were long established, Proterozoic marginal tectonic processes (p. 2) had developed about each of them a circumferential belt of mountainous terrain. The orogenic structures of both these girdles were directed inward towards the centre of each land mass. The interior varied from hilly terrains to wide basin plains of continental deposition (Fig. 2(b)), and was a region of gentle epeirogenic movement, locally either up or down.

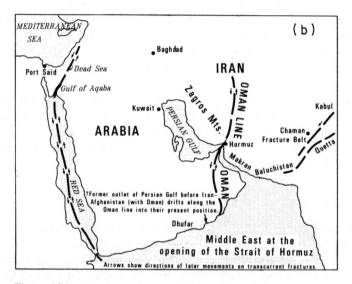

Figure 2(b)   Gondwana plates: Arabia, Iran, and India successively made the trans-Tethys journey to become attached to Laurasia. Arrows show directions of relative movement on the shears between them

Outside the two supercontinents, the globe was everywhere covered by sea. From east to west between the two continents stretched a mediterranean, the Tethys; elsewhere, and covering half the surface of the earth, stretched a pan-thalassa, the proto-Pacific Ocean. Either this was as large as all the present Pacific, Atlantic, Arctic and Indian Oceans taken together, or the globe was smaller during the Precambrian than it is now (p. 92).

During the late Palaeozoic both supercontinents moved majestically from east to west around the globe by continental drift. Gondwanaland in high southern latitudes actually moved across the south polar region so that its sedimentary rock system for that time (Permo-Carboniferous) includes huge accumulations of glacial debris (the Dwyka tillite) which is now found in all the southern continents (Fig. 2) and in India. Palaeoclimatic studies in several parts of Gondwanaland show by synthesis that the climatic girdles of the late Palaeozoic were the same in situation and possibly in extent as those around the Earth at the present time, and that the Earth's axis of rotation was presumably inclined to the ecliptic at the same angle ($23\frac{1}{2}°$).

At the same time the Laurasian supercontinent occupied a tropical situation as is evidenced (amongst other things) by the extensive zone of softwood, tropical coals reaching across Laurasia from Pennsylvania through Europe to China.*

Because at this stage of their history the two parent supercontinents often moved in sympathy, some authorities have tended to regard them as one continent and called it Pangaea; but this is a misconception. Each was a physical entity and each had its own girdle of mountains, folded towards its own centre. Their respective geologies differ fundamentally too: the Caledonian and Hercynian tectonics typical of Laurasia did not affect Gondwana much, and changes of sedimentology in the two supercontinents are often made in contrasted sense.

The passage from Palaeozoic to Mesozoic is marked in Laurasia, for instance, with new faunas and considerable making of horsts and grabens. In Gondwanaland it passed almost without notice: with continuous sedimentation and no tectonic activity. Even faunas and floras developed with continuity in a slowly changing topography.

Continental collision

Early in the Permian period Gondwana and Laurasia collided! The point of impact was in the vicinity of Senegal on the one side, and the Bahamas on the other (Fig. 3(a)). In Senegal a deep basin was subsequently formed, to be filled later with more than 1000 metres of marine Jurassic sediments of types not formerly evident in that area. The Bahamas is, alternatively, a submarine plateau of uncertain origin. These regions are shown in proximity on a continental reassembly by Bullard (1975); and evidence for it comes from stratigraphic distributions near Cape Hatteras and Cap Blanco recorded by Sougy (1962), by Rona (1970) and by Ferm (1974), following whom the matter was summed up by le Pichon et al. (1977). Briefly, a comparison is found between the early to mid-Palaeozoic formations and structures of the Appalachians in eastern North America (which are well known) with a similar set of folded formations (but lacking Devonian) along the coastlands of Africa from Senegal to Morocco (Fig. 4).

These Mauritanian formations, geosynclinal in origin like the Appalachians, are not related to the Atlas and anti-Atlas mountains of Africa which strike ESE–WNW, but to a chain of depressions — Essaouira, Aaiún and Senegal — with southwest to northeast trend (Fig. 4). The formations were

* Coals which succeed the glacial deposits of Gondwanaland are quite different. Their trees were deciduous and have strongly marked annual growth rings attesting a contemporaneous cold temperate climate as Gondwanaland at that time moved northward out of the south polar region. Contemporaneously the tropical coals of Laurasia were succeeded by the red Permo-Triassic desert sandstones as the continent moved northward past the Tropic of Cancer.

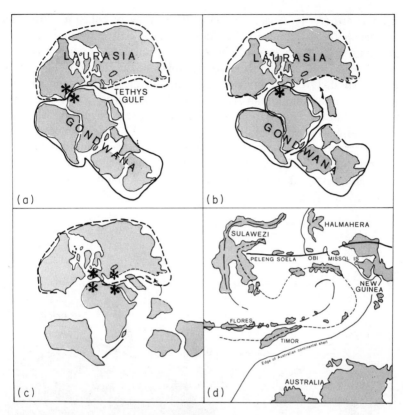

Figure 3 The glancing collision between Gondwana and Laurasia (with the Tethyan Gulf between) through four successive stages: (a) Permian impingement (**) between northwest Africa and eastern North America; (b) mid-Jurassic, the supercontinents rotate upon each other (Laurasia clockwise, Gondwana anticlockwise) so that the fulcrum of contact moves towards Gibraltar. The Gondwana fragments first of Arabia and then of Iran–Afghanistan which began to move out in the Permian are now well on the way to join Laurasia; (c) stages 3 and 4 of the collision. Drifting apart of the Gondwana daughter continents with opening up of the Central, North and South Atlantic (mid-Jurassic, late Cretaceous and early Cretaceous respectively), and Indian Oceans (early to mid-Cretaceous). Iran and Afghanistan have joined Laurasia, and India is on the way. The mountain systems of southern Laurasia and North Africa are impacted. Later, when Africa withdrew southward again, it left much of its mountain girdle north of the mediterranean in Italy, Yugoslavia, and Greece. The Falkland Islands–Agulhas fracture zone controls the separation of South America and South Africa; (d) the last phase of the collision. The intussusception of the Banda arc at Sulawezi by New Guinea (early Miocene). Further conflict is reported between Timor and the Australia–New Guinea mass

8

thrust landward (on to Africa) perhaps at the time of collision. Two facies are distinct, a folded zone on the landward side and a strongly deformed metamorphic zone to seaward. These latter disappear beneath thickening wedges of Mesozoic or Cenozoic strata which cover the coastal plain and continental shelf. Orogenic episodes recorded in the Mauritanian strata are described as 'mild Caledonian' (indicated by an hiatus) and strong 'post-Devonian within the Palaeozoic (Hercynian)'. These typical Laurasian orogenies are foreign to Gondwana; so here according to the evidence of stratigraphy and structure, is a sliver of North America that was left, after the impact, attached to Africa when the fulcrum of rolling contact moved farther northeast and eastward — opening up the North Atlantic basin in the rear as it did so.

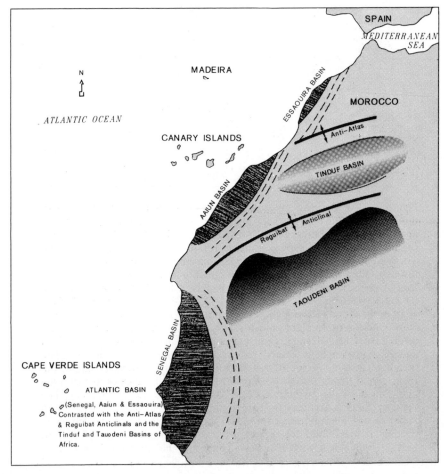

Figure 4   Discordant relation of the North American-type formations and structures of the Senegal, Aaiun, and Essaouira basins in Morocco with the more easterly trend of the Atlas structures. For the significance of these structures see text

After the collision, the Tethys strait which had previously existed between the two supercontinents was closed at the western end, affording opportunity for typical Gondwana floras (*Glossopteris*) and land faunas to migrate to Siberia and China where they now appear as fossils. Tethys became a long east–west gulf in which terrestrial detritus ranging in age from Jurassic to mid-Cenozoic was laid from both northern and southern margins. The waters were warm and teemed with pelagic biota so that marine petroliferous limestones were formed in consequence.

Impelled by their momentum, the two supercontinents then rotated further upon each other, Laurasia clockwise, Gondwana anticlockwise (Fig. 3(b)); and as the fulcrum of contact between them shifted northeastward past Casablanca, North America and Africa separated and the North Atlantic rift began. As a result of the ovoid shapes of the two supercontinents, the moving fulcrum caused regions in front of it to close and regions behind to open.

The full brunt of the collision was then felt as the fulcrum of contact passed Gibraltar (Fig. 3(c)). Africa, no longer in contact with North America, advanced northwards and closed the intercontinental Tethys sea which, being no longer connected with the outer ocean, was reduced to a group of basins some of which dried up into salt marshes by mid-Cenozoic time (p. 171). Although the modern Strait of Gibraltar opened up again later along the Azores–Gibraltar fracture zone, when the basins refilled to form the Mediterranean Sea, the eastern end still remained closed and the present situation of the fulcrum of contact between Gondwana and Laurasia seems to have been long established in or around Armenia.

So, during the time when the fulcrum was located between Gibraltar and Turkey, the full mass of Africa crunched into southern Europe progressively from west to east. The tectonics of this encounter were stupendous, and when (as the fulcrum passed eastwards) the Mediterranean opened up again and the continents drew apart once more the mountain girdle of North Africa between Tunis and Turkey was missing! At the crunch it had been welded on to Laurasia. The Tunisian mountain system is now continued through Sicily, Calabria and the southern Appennines (Cairé, 1971) where it was broken again. The continuation is found in the Dinaric Alps of Yugoslavia south of the Pusteria Line, and thence through the mountains of Greece to Crete and the Taurus Mountains of southeastern Turkey. All these ranges are rotated 35–50° anticlockwise (still recorded by their palaeomagnetic orientations). They are best regarded as a number of small plates torn off Gondwana and ground anticlockwise between the two supercontinents, while retaining their original (Gondwana) southerly folding.

By contrast, the Laurasian mountains from the Alps to the Caucasus are folded northwards on to that supercontinent. But these too are shown by palaeomagnetic observations to have undergone rotational and translational motions relative to the stable regions of northern Europe during Cretaceous and Eocene time.

The last remaining part of the Tethys is the Mediterranean, which had an

10

interesting history after it was closed at both ends (p. 171). A detailed account of the 'collision tectonics' in the Mediterranean region has been given, in the terminology of plate tectonics theory (p. 112), by Dewey *et al.* (1973). So through Turkey to Armenia the two mountain systems lie 'back to back' near the line of the late Cenozoic megashear of North Anatolia (Sengör, 1979). Younger displacements of this kind often follow the structural grain of older tectonic events, on continents as well as in the ocean basins (Windley, 1979). The region is still one of intense compression, strong seismicity and tectonic disquiet (see Cohen, Schamel and Boyd-Kaygi, 1980).

Eastward of Armenia collisions between Laurasia and Gondwana are of different kind. Laurasia remains there a single land mass; but three fragments of Gondwana (Iran–Afghanistan, India and Australia) have driven into Laurasia independently and at different times. This is because in late Jurassic and early Cretaceous time Gondwanaland fragmented into the five southern continents — Africa, South America, India, Australia and Antarctica — all of which drifted centrifugally apart as though the parent supercontinent had 'exploded' (Fig. 2).

The first fraction of Gondwanaland to break away was Arabia (Fig. 2), which later reattached itself to Africa as it is now. These events probably occurred in early or pre-Permian time.

Iran–Afghanistan began to leave East Africa in Permian time and collided early with Laurasia (Figs 3(b), (c)). The arid central plateaux of all these areas display vertical tectonics (epeirogenesis) with relatively uniform, epicontinental deposits. The regime also allowed for broad marine transgressions, and migration of troughs. There are no records of Caledonian or Hercynian orogenic disturbances such as afflicted Laurasia.

Details of the union of the Iranian and the Arabian structures, and the evolution of southern Tethys, are given by Stoneley (1981) from whom we learn that contact was made during the mid-Cretaceous and that the main collisions (resulting in the Zagros orogeny) was delayed until the Miocene.

Finally, Alpine–Himalayan mountain-making raised the weals of the Zagros, Mekran, Baluchistan and Sulaiman Ranges which have welded southwest Asia into its present unity.

The raft of peninsular India did not impinge upon Laurasia until later, and palaeomagnetic data from the Deccan traps suggest that at the time of their eruption (late Cretaceous–Eocene), peninsular India was still in the southern hemisphere. Several orogenic phases of the collision are determinable. The first orogenic impulse of the collision, manifest in the Karakorum, is likewise of late Cretaceous–Eocene date. The effects are spread along the Indus suture line with northwardly-directed folds and thrusts (Laurasian) in the Kailas Range and southwardly-directed structures in the Tibetan and higher Himalayas. Along the suture line itself are numerous blocks of exotic rock types of Tethyan origin and large masses of ultrabasic rocks. Sedimentary rocks stand steeply between vertical fractures typical of an orogenic 'root zone' (Gansser, 1974).

11

Cenozoic thrust movements in the Himalaya are all directed south, towards India. Thus on the Main Central Thrust the higher Himalaya override the Palaeozoic (Tethyan) rocks of Spiti and the structures of the lower Himalaya. In turn these are thrust southwards against the Siwaliks and slices of the Gondwana formations at the Main Boundary Fault. Lastly the Siwalik foothills have been carried southwards over the craton of India which has been bent down to form the Punjabi trough.

The northern half of the raft of India has thus driven beneath Tibet so that, uniquely upon Earth, the continental crust is here of double continental thickness, now perhaps 80–110 km thick in the western Himalayas (Menke and Jacob, 1976). Related earthquakes still continue. But palaeomagnetic data from Russia (Creer, 1970) suggest the Laurasian land mass is more stolid and 'has not undergone any significant continental drift since late Cretaceous times'.

The last item of the great collision, by Australia–New Guinea into southeast Asia, did not occur until the Miocene and took place then with strong Gondwana anticlockwise rotation so that the impingement drove in the Banda arc at Sulawezi (Fig. 3(d)). Evidence from deep-sea boring indicates that the Tasman Sea, between Australia and New Zealand, formed between 60 and 80 million years ago, and that the separation of Australia and Antarctica took place somewhat later (p. 161).

Laurasia was less active than Gondwana. It split into only two pieces, North America and Eurasia, although there are signs of a one-time incipient split through Siberia which was not completed.

The break-up of Gondwanaland

During the Palaeozoic and early Mesozoic eras both Gondwana and Laurasia drifted through global climatic zones which were then distributed over the Earth almost exactly as they are now. In Gondwanaland particularly, sedimentary rocks were laid down abundantly in interior continental basins, and these sediments record that during late Palaeozoic time the huge land mass drifted from east to west* across the southern polar region with appropriate climatic changes occurring in opposite sense on those parts approaching, and those parts retreating from, the pole respectively. For instance, the typical late Palaeozoic polar tillite deposits are oldest in South America and youngest in Australia; and the glaciation was over and done with in South America before it was begun in Australia (King, 1962/7, Table III). Other palaeoclimatic regions record conformable data.

In the mid-Mesozoic, quite different motions, and a spectacular series of disruptions, supervened. The twin supercontinents were dismembered, and their component parts (the modern continents) were drifted far away from

---

* With reference to the reconstruction, where South America appears on the west. The actual path over the globe, of course, was south towards the pole and then north away from it on the far side, without changing the directional movement of drift.

12

each other. Between these daughter continents opened up new ocean basins: the Atlantic, Arctic and Indian Ocean basins. A portion of the periphery of each continent was a segment of the original circumvallation of Gondwana or Laurasia, the rest was new, fractured coast. The first marine sedimentary formations along these new coasts serve to date their origins. Most are early or mid Cretaceous, a few are lower or mid-Jurassic. At this time the old minor split down East Africa which had admitted a Permian gulf from the Tethys as far south as the middle of Malagasy began to grow southwards once more, mid-Jurassic sediments were laid along the coast of Tanzania and by the beginning of the Cretaceous the gulf extended to the Cape, splitting Gondwanaland in two. By the middle of the Cretaceous both western and eastern halves had split again into the component continents much as they are known today. All the continental fragments drifted centrifugally and rapidly apart: South America to the west, Antarctica south, Africa nudged north so that it impinged on Europe.

Iran, already travelling north-northeastward, had not far to go before its way also was barred by Laurasia (King, 1973) to which it has remained attached ever since. Much of the country is covered with terrestrial accumulations of Cretaceous or Cenozoic age, but sufficient of the older rocks is exposed for us to be assured that they are of Gondwana type.* Above the basement the Palaeozoic rocks are shallow-water marine, lagoonal and continental in type, with many hiatuses, while Caledonian and Variscan (Laurasian) tectonics are not known (Stocklin, 1964, 1974). Iran could have assumed its position relative to neighbouring continental blocks early in the Mesozoic.

A curious relic of the old Tethyan sea remains north of Albourz (a part of the Gondwana margin); the southern half of the Caspian Sea is still floored with oceanic basalt (Stocklin, 1974).

The geology of the Indian peninsula is typical of Gondwanaland. The Himalaya in the north is overfolded towards the peninsula in the south and is part of the circumvallation of Gondwanaland. The question is, how far to the north does the Indian land mass extend? It appears to be thrust under the land mass of Laurasia far beyond the crystalline rocks of the Brahmaputra (Tsangpo) valley. The mountain structures of Kuen Lun are folded northward and are therefore Laurasian. Several observers (e.g., Stoneley, 1974) have noted the abnormally high plateau of Tibet, and geophysics has recorded that the thickness of continental crust beneath the whole plateau is twice normal as though considerable underthrusting of the northern land mass had extended far beneath southern Asia. Geological evidence indicates that these tectonic relationships developed during the Cretaceous period.

* Typical of many letters I have received, Dr J. W. Schroeder writes: 'Due to similarities between the Salt Range and the Hormuz it was natural for me, 30 years ago! to place India almost adjacent to Arabia. But if Iran, as you suggest, is placed between Arabia and India, then the Hormuz series should also be present in Central Iran: this series is *indeed* present (in the Lut block!).'

Of all the Gondwana fragments, Australia travelled the farthest. It went eastward (with 90° anticlockwise rotation) until its frontal margin (now New Zealand) was far advanced into the Pacific and the marine mid-Cretaceous fossil faunas show a change from earlier Tethyan types to Indo-Pacific. Then, as Antarctica withdrew south it provided a temporary, late Cretaceous, coastwise connection between Tierra del Fuego and New Zealand. Patagonian mollusca took the opportunity to invade New Zealand shores so successfully that, to this day, the evolved molluscan fauna of the antipodes is largely descended from the late Cretaceous Patagonian immigrants!

Curiously, at the point farthest east, the direction of Australian drift changed to northwest. New Zealand, with the New Caledonia and Lord Howe submarine ridges, was left behind and the Tasman Sea opened, as Australia and New Guinea (the latter with its long westward-projecting ridge on which stand the islands of Misool, Sula, Obi and Peleng in the van) drove towards southeast Asia (Fig. 3(d)). The continental encounter (of Mio-Pliocene date) was truly aimed: the long lance penetrated the Banda arc, and turned the structural trends of Sulawezi inside out. This continental joust has recently been the subject of stratigraphical analysis by Carter, Audley-Charles and Barber (1976) and Carey (1976). It forms a worthy conclusion to the complex collision of two super land masses, Laurasia and Gondwana.

Most post-Palaeozoic continental drift seems, on geological evidence, to have been accomplished during the Mesozoic era. Only small episodes of drift have been dated as (a) Mio-Pliocene, or (b) Pleistocene, and over most of the Earth the early Cenozoic was particularly free of such events. So it was from the late Mesozoic mayhem that the modern pattern of continents and ocean basins emerged.

But the usual interpretation of tectonic activity involving orogenesis, volcanism and the world's major seismic patterns which is facilely described as 'circum-Pacific with a long zone extending from southern Europe to Indonesia' lacks insight. The Palaeozoic oceans comprised (a) the Tethys, a long narrow sea between Laurasia and Gondwanaland, and (b) a proto-Pacific. These together surrounded both the supercontinents. When the break-up of the parent land masses occurred, the new (modern) continents dispersed in all directions. Some between Afghanistan and Sumatra drifted across the Tethys and attached themselves to the opposite continent so that Tethys itself survives only as a shrunken Mediterranean. Others went out into the proto-Pacific and now appear as a ring surrounding a diminished Pacific Basin (which is nevertheless the largest physical feature on Earth). The present girdle of orogenic activity about the Pacific Ocean is only a convergence of fragments of the older circumvallations of Gondwana and of Laurasia. All the late Cretaceous to Recent tectonic activity (including mountain-making, volcanism and high seismic activity) is no more than continued activity on the old sites, the active marginal zones of the two Palaeozoic supercontinents. Gondwanaland and Laurasia, though dismembered, are still with us and are still functioning, as any reader of *The Geology of the Continental Margins* (Drake and Burk, 1974) will find.

Chapter 2

Tectonic activity on a spherical Earth

Tectonics of a spherical Earth

The outer Earth is a relatively thin (max. 100 km) *crust*, rigid above, ductile below, over a hot, mobile interior of peridotite, called the *mantle*. The upper part of the mantle (sometimes called the asthenosphere) is especially active, being the zone where large quantities of volatile constituents discharging from the mantle below are prevented from escaping freely into the atmosphere (and hydrosphere) by the solid crust above. Much tectonic activity originates in this ebullient zone which extends beneath both the ocean beds and the lands alike.

Heat flow rates from beneath both continents and oceans are nearly equal.

Throughout its volume the mantle deforms, apparently, by plastic viscous flow. The asthenosphere (sometimes called the 'low velocity layer' because earthquake waves are transmitted more slowly therein) operates in many areas by more fluid viscous flow, so that it approximates a hydrostatic geoid. Above this level, continental basements prove by the abundance of Precambrian gneisses that ductile laminar flow is general within the lower crust, with minor structures often induced by 'ductility contrasts' between individual mineral fragments. Many granites here 'were finally intruded as concentrated crystal mushes or even in the solid state'; and 'the construction of a composite batholith takes some 50 to 70 million years' (Pitcher, 1975). Beneath the ocean, the transition is usually from subjacent basaltic magma to pillow lavas and flows of crustal basalt. The rigid oceanic crust deforms shallowly by fracture and the lateral displacement of large masses takes place apparently with minimum internal deformation.

The force of *gravity*, operating everywhere in sectors of the Earth body from the surface towards the centre, tends to draw materials towards the core of the planet and to densify those materials in the process. Its operation is normally *down*, and it produces by density a radially layered Earth-structure (p. 88). Additionally, gravity operating downward within a sphere

15

generates at the surface a small but ubiquitous horizontal force of lateral compression which pervades the upper crust, as was found by Hast (1969).

Conversely, the little energy demons of mantle activity seek to get out. From the interior of a sphere, there is only one way out: it is *up* towards the surface. Hence the active tectonics of the Earth are expressed radially (or at the surface, *vertically*); Magmas and plutons ascend into the crust, ridges and domes arise upon the sea floor and upon the continents, and where the mantle (or the asthenosphere) is rich in volatile constituents, volcanic peaks built of levitated rock-material rise the highest of all. Recall, as the result of subcrustal, mantle activity, the summit volcanoes of the Andes; Mount Elbruz, the highest mountain in Europe, a volcano surmounting the Caucasus; Hawaii and thousands of other volcanic peaks rising from the deep floor of the Pacific Ocean.

So the fundamental tectonic mechanisms of global geology are *vertical, up or down*: and the normal and most general tectonic structures in the crust are also vertically disposed. But when we write vertical on a sphere, we mean *radial*, and so for an Earth sphere we really mean expansional. This is a quality, not of geological control, but of spherical geometry (p. 88).

Both these processes were manifest early in the history of the Earth, and as a consequence the materials of the outer Earth became segregated in spherical layers according to density. By 2500 million years ago most of that major segregation was accomplished so that (although the geographical distribution of land masses has changed greatly through the geological ages) the total volume of continental matter then produced was similar to what it is today. Since that time there has been 'neither net accretion nor retreat of the continental margins on a global scale'. Instead, the continental margins should be regarded (Wise, 1974) as the operational focus of a 'constant volume system serving to weld escaping sediments (supplied by denudation) back on to continental rafts of constant area and constant thickness with possible shifts in the equilibrium level modifying the details of the rewelding process'.

Beneath the crust, 52 per cent of the Earth, the mantle, is of basic and ultrabasic rock types (peridotite) hot and charged with lighter volatile substances — largely superheated steam and carbon dioxide.

The answers to all tectonic problems at or near the Earth's surface should therefore be sought *firstly* under *vertical* explanations and methods. Through the geologic ages, the Earth's crust has operated like a heavy blanket over an uneasy sleeper, heaving up in domes and arches at some places and subsiding into hollows and troughs elsewhere. These directions of local movement have sometimes been reversed. Sedimentational troughs, for instance, have quite normally been later erected into mountain ranges. Elevation and depression of the crust are universal, and as the source of heat energy is radioactive decay within the mantle, they are likely to continue so for a very long time.

The relatively thin, rigid lithosphere lacks great strength and is not capable of maintaining itself without deformation over horizontal distances much exceeding 400–500 km. A mountain range representing an excess load upon the crust may be thousands of kilometres long, but if the width of the range exceeds the critical limit of about 400 km the crust cannot maintain the load of the mountains by rigid strength alone (Gunn, 1949). It fails, sags and sinks vertically into the mantle below, so transferring some of the load to the subcrust. Conversely, as mountain systems become lighter by denudation they tend to rise. So the lithosphere maintains, within the strictures of internal friction, a balanced (isostatic) state.

The changes of level sometimes attain differences exceeding ten kilometres and the corresponding changes of load are considerable. By what mechanisms within the crust are these alterations of elevation and balance achieved? Though different kinds of deformation are known near the surface, ranging from brittle fracture to fluid flow according to the local state of crustal rigidity, the process which is dominant in depth appears to be steep laminar rock creep: which permits thousands of metres of differential vertical displacement with the formation of schistose and gneissose rock types and structures (Fig. 6) at depth, combined (at first sight) with negligible horizontal stress or displacement.

Self-induced horizontal tensions or compressions are generally associated with upward and with downward displacements respectively, and are due, of course, to centripetal gravity operating within the sphericity of the Earth.

The central part of the Earth is a solid, metallic *core* of iron–nickel. Occupying 16.2% of the Earth's volume, its density ranges from 11 to 16.

Cymatogens

Cymatogeny (King, 1962/7) (Fig. 6) is a mode of major vertical deformation of the continental crust wherein the fundamental structure induced at intermediate depths is a steeply inclined tectonic gneiss, with or without minor igneous activity, which is usually basic only. At the Earth's surface it is expressed by arching, scores of metres in height, hundreds of kilometres in width, and sometimes thousands of kilometres long. Rift valleys due to tension (induced in the rigid crust by the arc–chord relationship), and wedge uplifts due to unbalanced forces, are frequent minor attributes of the outermost crust.

It is the arching, however, which is the most important, and the most expressive description of cymatogeny is perhaps 'the undulating ogeny' (Greek κυμα = wave). Some arches are formidable (the Andes), some are inappreciable to the eye (the Congo–Zambezi divide).

Upon exaggeration cymatogeny passes over into orogeny.

Cymatogeny has been the characteristic tectonic activity of late Cenozoic and Quaternary time, and its effects are apparent now in the most striking scenery of our planet. Its primary source of energy is still to be sought —

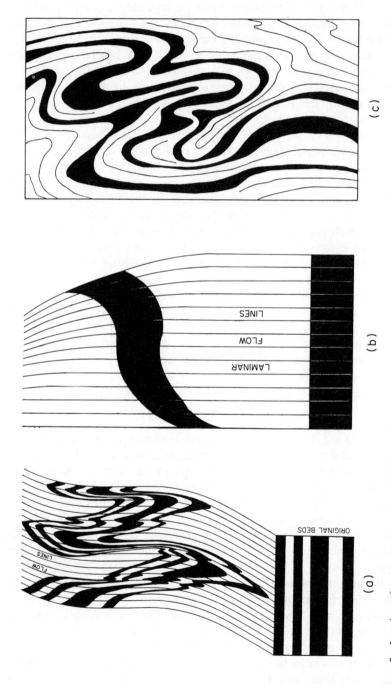

Figure 5 Laminar flow in rock masses: (a) the thickness of any formation along parallel laminae remains constant; (b) where laminae converge the volumes between laminae remain constant; (c) laminar flow in an oil-slick shows how extremely attenuated folds may develop (after S. W. Carey)

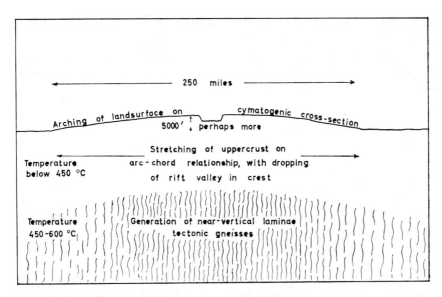

Figure 6   Diagrammatic crustal section through a typical continental cymatogen relating major arching and rifting of the surface to formation of tectonic gneisses (often Archean) at intermediate crustal depth. Displacements are in the vertical sense only. At greater depth levitated mantle may be the active agent of the cymatogen (Fig. 13)

below, in the asthenosphere or upper mantle. This is apparent from the structure of sea-floor cymatogens (Fig. 13).

Typical cymatogens may be quoted from every part of the world: the belts of country affected by the arching and doming are major features of the landscape. Vertical uplift often exceeds 1000 metres. Yet rock deformation near the surface is at a minimum. Only in the deeper, hotter zones do metamorphic transformations of material almost *in situ* form schists and gneisses. Milder metamorphism beneath the flanks of larger cymatogens has caused three zones of: (a) zeolite production; (b) greenschist; and (c) amphibolite; to appear in descending order.

Despite widespread belief to the contrary, there is little geological evidence associating cymatogens with large horizontal movements of the kind expressed in continental drift and sea-floor spreading, although such features may sometimes be fortuitously associated.

Sometimes new cymatogens have rejuvenated the topography in former orogenic zones which have been planed by denudation. Most of the world's orogenic ranges, folded in the mid-Cenozoic, were obliterated and the terrain reduced to a plain by Miocene and earlier erosion. Renewed cymatogeny later arched the Miocene plain into tablelands on the old mountain site. After this uplift, renewed valley incision carved the tablelands into a maze of later mountains, some of which preserve, as their summits, remnants of the Miocene planation. Others reveal, when viewed from similar altitudes, an 'accordance of summit levels' (King, 1976) (Chapter 10).

Cymatogens provide great variety of scenery ranging in form from simple topographic arches to complicated rifted and fault-blocked terrains developed through several successive tectonic phases.

Tilts, arches, and rift valleys

The simplest mode of deformation that a terrain can undergo is a tilt in one direction. Britain, for example, has repeatedly been uplifted in the west and depressed beneath the North Sea in the east (p. 194). The most recent movements are certified by the increased tidal reach of the Thames above London Bridge since Roman times, with increased danger of flooding the city. Cenozoic tilting is shown by contours on the early Cenozoic planation (Moorland surface) (p. 188), and many earlier movements in the same sense are attested by the geological map with the early Palaeozoic and older rock systems in the west and progressively younger rock systems cropping out towards the east. Offshore data from oil wells verify the greater depth of older rock systems beneath the North Sea.

Complex Cenozoic deformation has overtaken South Africa. The main upwarping parallel with the coastline, and 250 km or so inland, separates the uplifted, but basined, interior plateau with the Kalahari desert at its centre, from the outer, marginal monoclines that bend down beneath the sea (Fig. 7). Across this subcontinental deformation a number of upwarped axes (Fig. 8) run from northeast to southwest and these now constitute major divides (e.g. Kalahari–Rhodesia axis, Griqualand–Transvaal axis).

In North Africa, too, Lake Chad is an area of centripetal drainage not due to depression but surrounded by watersheds of greater elevation than the centre.

In these examples, deformation does not pass the stage of broad flexing. Rupture of crust and fault displacements do not normally appear: but the amount of local differential vertical displacement on the early Cenozoic (Moorland) planation may exceed 1000 metres. On the Mesozoic (Gondwana) planation it is twice as much (p. 189).

In East Africa also there are strong cymatogenic arches and domes (e.g. Kenya, Fig. 9) and these bear the most famous rift valley system in the world: but many observers, (e.g. Bailey Willis) who have set out to examine the rift valleys have ended up much more impressed by the greater crustal warping. For years controversy thrived as to whether the rifts were tensional or compressional in origin,and indeed, examples of each kind could be adduced. Nowadays the old difference of opinion is dead, and viewed cymatogenically, with laminar vertical deformation, the main rifts are seen to be extensional features on the arc–chord relationship across the topographic arch of the rigid upper crust, whereby crestal blocks have fractured and subsided under gravity (cf. Fig. 9). Unbalanced application of forces, or non-

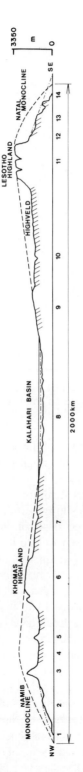

Figure 7 Cross-section of South Africa showing interior basin, rim highlands and monoclines of the coastal belt. The rim highlands display Jurassic and Cretaceous summit planations, and these planations flex downwards (dashed line) towards the interior basin where they are covered by late Cretaceous, Cenozoic, and Recent sediments of the Kalahari. In the coastal belts both east and west, these planations flex down in monoclines to pass beneath appropriate Cretaceous marine strata (Chapter 10). A later, early Cenozoic planation (dot-and-dash line and diagonal underlining) shows a similar differential uplift of lesser amount. At the coasts it passes beneath early Miocene sediments (see Chapter 10). 1, Atlantic littoral; 2, Namib; 3, Brandberg; 4, Omaruru Flats; 5, Erongo; 6, Windhoek; 7, Damaraland Plains; 8, Kalahari Basin; 9, Kaap Plateau; 10, Vaal River; 11, Lesotho; 12, Drakensberg; 13, Nottingham Road; 14, Durban

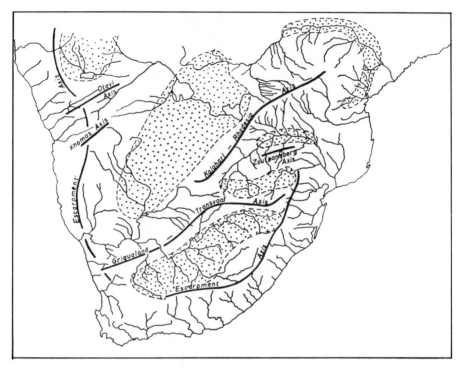

Figure 8   Crustal arches of the interior plateau of South Africa which govern the watersheds of the main rivers (modified from A. L. du Toit)

homogeneity of local geology, produces most of the secondary compressional features, of which the most outstanding is the mighty uplift of Ruwenzori in the very axis of the Albertine–Edward Rift.

In the deeper (ductile) crust and subcrust, there is universal compression (Hast, 1969), and displacements are only along vertical planes.

This is not the place to give a detailed description of the East–Central African rift system, the purpose is merely to place it among the prominent phenomena caused by *vertical* crustal movements on a sphere. Readers desiring further information will find, amid an enormous literature, useful papers by Pallister, Baker, Mohr and Whiteman in *Tectonics of Africa* (1971).

Certain rifts (e.g. Manyara, Eyasi) swing away from the main line and become one-sided. First Cloos and then Luchitsky (Fig. 10) showed that where the cymatogenic uplifts are domed the rifts bifurcate and the continuations become one-sided. The Leine Graben of West Germany shows symmetrical bifurcation at both ends. Where the dome is oblique to applied forces, rotational effects appear (Lake Baikal Rift, Fig. 11). The 'triple junctions' of rift valleys are not always symmetrical and there is often combination with pre-existing features. Much use has been made of these natural patterns to analyse the forces that were active at their genesis, e.g. Burke and Whiteman

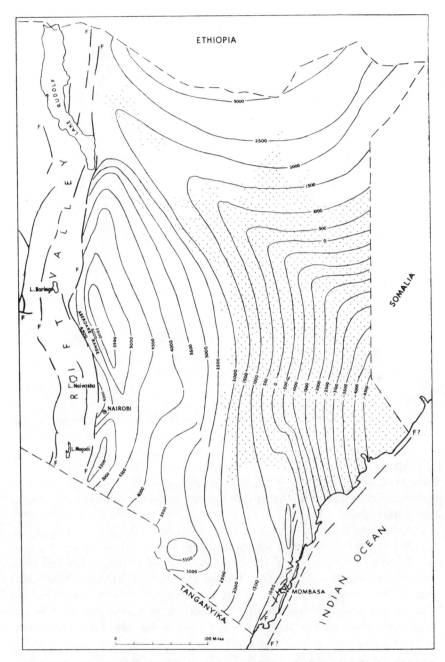

Figure 9   Contours (in feet) upon the early Cenozoic (Moorland) planation of Kenya showing its deformation during Miocene to Recent time. Area of Quaternary sediments stippled; major faults marked F (after Saggerson and Baker, 1965)

24

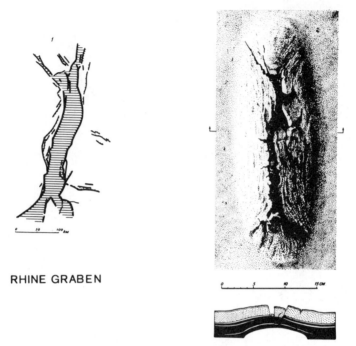

RHINE GRABEN

Figure 10   Bifurcation at the ends of a rift valley formed over a
domed uplift, Rhine Graben (after Luchitsky)

(1972) and Atwater (1970). These afford understanding of geomorphic evolution. More and more of these domal uplifts yield seismic evidence of intrusion at the base of the crust by masses of levitated mantle (7.1–7.5 km/s), e.g. East African Rifts and the African–Arabian dome. The rift faulting appears at the crest of the arch solely because the rigid crust comes under extension as the arch rises above the hydrostatic geoid, at which level vertical laminar flow is paramount.

Northeastern Brazil is a structural arch as shown by contours upon the base of terrestrial Cretaceous formations (Fig. 12). This arching was cumulative through several stages in the late Cenozoic, after the region had been smoothly planed by denudation during the early Cenozoic (Moorland) surface. Strong uplifts occurred at the beginning of both the Miocene and Pliocene periods, but probably the greatest deformation occurred in the Pleistocene when the cymatogen cracked open along fresh fault scarps near the crest of the arch to form the Sâo Francisco graben. A second rift valley in Brazil is followed by the Paraiba River. It separates the upcast and tilted Serra de Mantiquera on the north from a lower plateau upon the seaward side. Both Brazilian rifts lie parallel with their neighbouring coastlines.

The following is a rough list of well known rift valley structures: East African rift system extending to Jordan; Equadorian (high Andean); Sâo

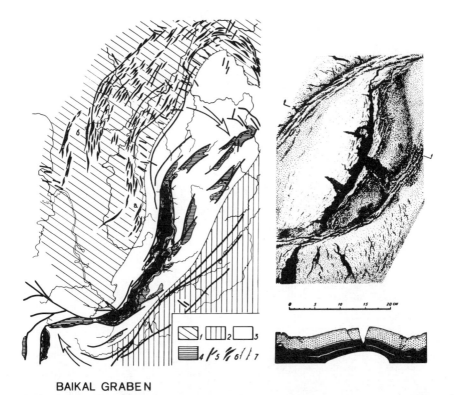

BAIKAL GRABEN

Figure 11   Baikal Graben, Siberia, formed with a left-handed wrench (after Florensov 1969 and Luchitsky)

Francisco and Paraiba rifts of Brazil; St Vincent and Spencer Gulf rifts in South Australia; George VI Sound and Lambert Glacier rifts in Antarctica; Rhone Valley and Rhine Valley; Oslo graben and Leine graben of Europe; Rocky Mountain trench and Rio Grande rift in North America; and Lake Baikal in Siberia. Almost all of these occur towards the crest of broad crustal arches, usually cymatogens. All except the Equadorian rift and the Lambert Glacier rift in the Antarctic have been visited by this author.

The highlands of southeastern Australia are also cymatogenic, with three special domed uplifts in the New England Range, the Blue Mountains and the Australian Alps separated by geocols at Muswellbrook and Goulburn respectively. No rift valleys are associated, unless the Cullarin scarp shows incipient rifting; but sporadic, mostly mid-Cenozoic basaltic outflows are associated from Victoria to Queensland.

Dome and basin cymatogeny, with associated faulting, has been characteristic of parts of the Rocky Mountain belt, e.g. Bighorn Mountains and Wyoming Basin; and the Rocky Mountain trench in Montana has been claimed by some observers as a rift valley. In New Mexico the Rio Grande rift, filled with 3000 m thickness of Pleistocene gravels, also qualifies; but

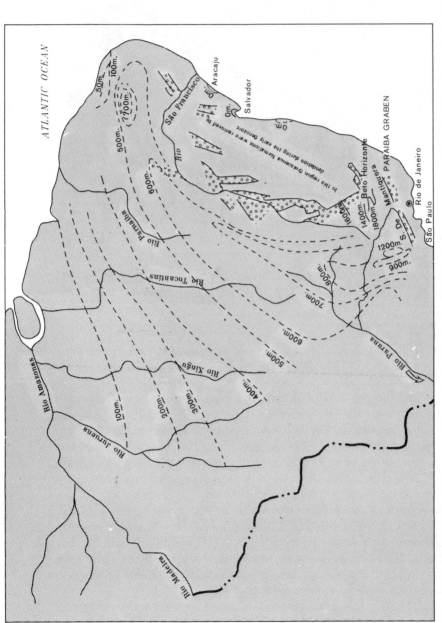

Figure 12   Crustal warping in Brazil. Contours (in metres) upon the base of the Cretaceous terrestrial rock series from the São Francisco Graben towards the Amazon River, showing the western side of a continental cymatogen. The eastern side comes more steeply down to the sea, and has been much eroded. The Paraiba Graben south of the Serra da Mantiquera is also shown. These grabens are Plio-Pleistocene in age

the region is structurally complex and the subjacent active cymatogen (a part of the East Pacific Ridge (p. 169) which is possibly young and simple) has to make its effects apparent through a continental crust already toughened by several orogenies ranging from Precambrian to Laramide (Cretaceous) in age. Geological descriptions of the Atlas Mountains too, quite rightly emphasize the phases of folding (including nappes) under which orogenic mountains were first raised upon the site. But those mountains disappeared under denudation during the early Cenozoic. Later deformation has been wholly cymatogenic. Cairé (1971) writes: 'Throughout all the Atlas ranges, the upper Miocene to Quaternary phases produce vertical deformations often oblique to the east–west direction of the previous structural zones.'

Few of the major mountain ranges of the Earth, indeed, were later than mid-Cenozoic in their orogenic (rock-deforming) phase, and planation of the original mountains was achieved almost everywhere before the Pliocene period (p. 188). These areas have been rejuvenated as highlands by later vertical tectonics and their present forms have been sculptured out of the late Cenozoic cymatogens by streams, rivers and glaciers. They are *denudational* mountains, and many of these modern mountain systems still preserve summit bevels and summit accordances testifying to this history (King, 1962/ 7, 1976). Thus at 4000 metres upon the Andes of Peru and Bolivia the Pliocene *altiplano* stands uplifted, and in Ecuador a typical rift valley has been discovered at the crest of the arch.

The mightiest of all mountain systems — the Himalaya — has yielded evidence, both in ridge–summit features and in the two series of sediments (Murree and Siwalik; Miocene and Pliocene respectively) on its southern flanks, of planation cycles that twice obliterated the mountain forms before the latest cymatogenic upheaval raised the parent (Pliocene) landscape for the glaciers and rivers to sculpt the present mountainous aspect. This history is very like that of the European Alps which also were orogenically formed towards the end of the Oligocene period, when all the main structures were completed. By the end of the Miocene the original mountain system had been reduced under denudation to a landscape of minor relief. This rolling landscape was cymatogenically re-elevated at the end of the Miocene, and was again reduced to a lowland by the end of the Pliocene when, for a third time, powerful vertical uplift carried that lowland to high elevation where it may still be identified as the 'alpine gipfelflur'. The present valley systems of the mountain scenery were excavated by glaciers and rivers only during late Pliocene and Quaternary time, thus isolating the present groups of peaks and ridges. Detailed studies of the present remnants of the gipfelflur indicate upon the major east–west cymatogenic arching of the whole alpine system several minor corrugations in the same direction, and these are crossed by transverse warpings. The major groups of alpine peaks are situated on the major domes where the two systems of warping intersect. Adkin (1949) has shown similar sets of longitudinal and transverse summit warpings in the Tararua Ranges of New Zealand.

Cymatogenic crustal arching appears also *before* the powerful rock-deformation (structural and metamorphic) which signals the onset of orogenesis; and indeed the main process of orogenic mountain-making appears nowadays to be a *vertical* function of the Earth's crust, activated from directly below the zone of deformation, where seismic and other geophysical data have in some instances indicated the presence of large bodies of rock material (either metamorphic or intrusive) of anomalously low density (p. 00). In either case by comparison with adjacent rocks the levity appears to be due to associated volatiles of which siliceous, superheated steam is usually the most prominent member (King, 1962/7).

This is a very different opinion from that stated by a previous generation of geologists, e.g. W. H. Bucher (1933) who ascribed mountain-making as due to 'crustal compression', and confidently postulated '250 km of crustal shortening' as responsible for the generation of the European Alps.

An opinion by van Bemmelen (1969) on the Alpine Loop of the Tethys Zone is confirmatory: 'The main conclusion of this geodynamic analysis is, *that in the case of arcuate mountain belts and island arcs, the direct cause of the orogenic evolution has to be sought in geodynamic processes at a meso-tectonic scale occurring in the upper mantle, directly underneath the mobile belts.'*

The more the subject is explored, the more examples arise where orogenic zones have repeatedly been rejuvenated, or where new mountainlands have risen into existence closely alongside former mountain systems. In some instances this can be demonstrated to have recurred from Precambrian until modern time. One such instance is Lake Baikal in central Siberia, which lies in a modern rift valley slashed across a great cymatogenic dome that appears to have formed under a wrench movement (Fig. 11). The present landscape is carved into valleys by the 300 rivers which sustain the lake; but the mountain summits around the lake show unmistakably an early Cenozoic planation which in Miocene time began to upcast to make the heights. These are the modern forms; but the history of crustal uplifts at the Baikal site goes back (Belitchenko and Khrenov, 1969) to Precambrian time: 'Thus the brief analysis of the geosynclinal stage of development shows convincingly that the Baikal mountain region . . . is one of the striking examples of the polycyclic development of the geosynclinal systems of the Earth.' Archean, lower middle and upper Proterozoic; early Caledonian; Mesozoic and Cenozoic tectonic episodes are quoted; and of the Cenozoic rift they state: 'Although the central deep rift is about 50 km wide, sub-parallel rift faults extend in a zone some 150 to 200 km wide. The faults at surface dip normally at 60° to 70° into the grabens, but the structure is essentially vertical.' Florensov (1969) considers that beneath the entire rift zone lies a crust–mantle mixture extending up into the lithosphere. Indeed, seismic refraction velocities indicate the presence beneath Lake Baikal of slivers of oceanic-type crust such as are present in the much younger Red Sea, the Gulf of California, and the southern Caspian basin.

Baikal surely is a cymatogen *par excellence*; and its long continuity upon the same site is an outstanding proof of vertical tectonics within the Siberian crust.

Recent seismic results obtained on the western flank of the Gregory Rift Valley in Kenya (Maguire and Long, 1976; Long and Backhouse, 1976) showed the continental crust there to be underlain by an ellipsoidal body of anomalous mantle (7.1–7.5 km/s) which created the Kenya Dome. From this subjacent mass, more than 100 km depth, arises a steep intrusion of the same levitated mantle which occupies a split in the lithosphere precisely below the rift valley.

Sea-floor cymatogens and rifts

From the foregoing reviews of continental rift systems the cardinal principle which has emerged is that the activation of continental rift valleys has nothing to do with general states of tension or compression in the Earth's crust (although local tensions or compressions may sometimes be generated thereby). Instead, the tectonics involved are vertical in expression and are activated by the accumulation of levitated (gas-impregnated) upper mantle rock that rises and is injected, probably up a preexisting megashear, into the lower crust (Fig. 13).

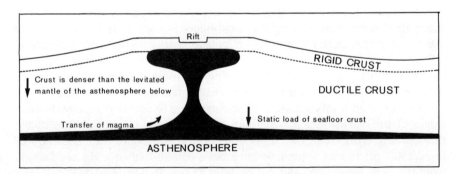

Figure 13 Section through the central part of a sea-floor cymatogen to show the underlying body of levitated mantle rising (like a salt structure) from the asthenosphere into the overlying suboceanic crust (after M. Ewing, 1965)

Where this mechanism exists, crustal activity and doming follow. If the body of levitated mantle is linear in distribution, as by streaming within the mantle, the related phenomena in the upper crust are zonal upwarps. *Except where megashears are present, no association with large-scale horizontal displacements of the crust such as continental drift is necessary.*

Remote-sensing techniques used within the ocean basins of recent years have not only indicated the presence of ocean-floor rises and ridges, sometimes with a crestal rift, but have also indicated the presence beneath them of a typical body of levitated mantle, with associated high seismicity at shallow (crustal) depth, a narrow zone of enhanced heat flow and occasional basic or ultrabasic volcanism. The ocean floor being of simpler geological structure than the continents, these results are even more clearly displayed in the ocean depths than on the lands.

For all the oceans, the new knowledge has been summarized and expressed pictorially by Heezen and Tharp in a series of diagrams that is deservedly famous and is summarized for the globe in the beautiful chart of *The Floor of the Ocean* published by the American Geographical Society with the support of the US Navy Office of Naval Research.

These diagrams bring out one curious difference between cymatogens on land and at sea. The arches may be followed continuously upon the lands for great distances, the sea-floor cymatogens are deemed to be crossed by large numbers of cross-fractures which usually make a marked angle with the direction of the sea-floor ridge. The ridges of the southern oceans show this right up to 45° angle (Figs 14, 15, 42). The crestal rift is thus divided into short sections, and these sections are offset with respect to one another. A similar arrangement of geological features is found where, as the geologist would say, the strata have been crossed by two systems of faulting (or jointing) at an angle to each other. These divide the rocks into a series of lozenge-shaped blocks with offsets like those on maps of the submarine ridges.*

It is surmised, therefore, that some such shear structure is present in the submarine ridges. Einarsson (1968) has indeed demonstrated this pattern of shear in the Reykjanes Peninsula of Iceland, and believes it to continue in the Reykjanes ridge to the southwest, on the sea floor of the mid-Atlantic, and that a criss-cross pattern of fracturing has rotated the position of blocks of the ocean-floor crust bearing the rift valley features at the surface.

So widespread is this apparent offsetting along the cymatogenic mid-ocean ridges of the world, that it is perhaps a fundamental shear structure of the thin basaltic crust (see Hast, 1969 and p. 188). Certainly the offsettings of the crestal rift blocks appear to have been established in a zig-zag fashion from the very beginning, and not by later re-orientation (Fig. 14).

The forces we have discussed so far are expressed *vertically* by gravity and anti-gravity (or gas-levitation), whether beneath the lands or below the sea.

A further problem then appears. Horizontal displacements about the Earth's surface, exemplified by continental drift and megashears (upon both the lands and the sea floors) require tectonic explanation. Either of two interpretations becomes possible: the first (expressed by the geophysicists' credo of the foreword) derives from the observation that sea-floor ridges often (though not invariably) bear rift valleys at the crest, and proceeds to

---

* In the coastal hinterland of Natal, South Africa, fault zones have shattered shales of the Ecca series. Though no deterioration may be immediately visible, quarried blocks which have been left exposed for a few months weather so rapidly along the intersecting joint planes that an apparently sound block crumbles to a pile of small lozenge-shaped pieces at a single kick.

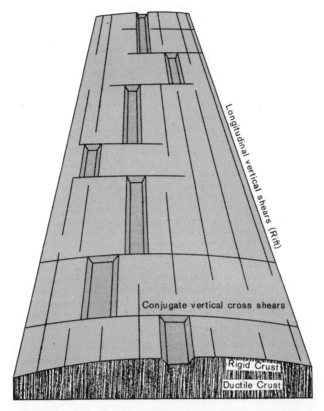

Figure 14 Structure of a cymatogen in thin, rigid upper-crust based on vertical uplifts along two parallel systems of parallel shears which intersect orthogonally. There are no horizontal displacements of the longitudinal rift on the conjugate fractures; but the displacement changes from one longitudinal fracture line to another

argue that such rifts demonstrate a state of extensile stress acting at right angles to the orientation of the ridge. With the injection of basic rocks along the crestal zone new material is added to the submarine crust which then 'spreads' to either side away from the centre line of the ridge. The structure which subsequently develops would include successively older basal marine strata away to either side. Adjacent land masses must therefore move aside, or the sea-floor towards the continental block must be subducted (Chapter 4).

In support of this 'sea-floor spreading' are quoted stripes of geomagnetic reversals that are present on the flanks of sea-floor ridges and across the ocean basins (Chapter 6).

To aid the interpretation, a new type of fault (*transform* faults), which can only exist if there is crustal displacement, was defined by Tuzo Wilson.

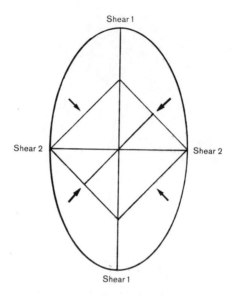

Figure 15 Stress ellipsoid for a region of universal but unequal compressive stress. Bold arrows show maximum horizontal compression. Shear directions: 1 = maximum shear; 2 = conjugate shear. According to Hast (1969): 'Horizontal compressive stresses in the bedrock were found at all points of measure on Iceland. . . . Where the Mid-Atlantic Ridge is assumed to encounter Iceland, the orientation of the maximum horizontal shear in the Icelandic bedrock agrees with the orientation of the ridge; this suggests that the ridge may have its origin in shear stresses in the crust. The maximum horizontal pressure acts in a direction at 45° to the ridge, instead of the 90° direction to be expected, where new material extruded to form the Atlantic Ocean floor'

In simple instances, theories that the rift faulting is of horizontional, extensional nature may express the whole truth; but most rift valleys are only minor landforms induced at the crest of a much larger crustal arch. This fundamental arch, or dome, or suboceanic ridge is normally (Hast, 1969) under compression by a concentric force of gravity acting within a sphere. Under these conditions the arch comes under shear both longitudinally and across (Chapter 7).

The whole subject of lateral (horizontal) activity builds up in the 'plate tectonics' hypothesis (Chapter 7). For this purpose a true assessment of the spherical nature of the Earth's surface is gained by the use of Euler's theorem, which states that any motion about a spherical surface may be expressed as a rotation about an axis through that sphere. Much advance has been made in geotectonics by this method of thought and comparison (Chapter 7).

Beginning with these original data, the second interpretation regards the sea-floor ridge as an arched uplift, vertically controlled, and the crestal rift as a superficial feature only, which is produced by stretching of the brittle upper crust over the arch. It appears naturally between the two oppositely tilted flanks. Flat and wide sea-floor ridges, like the southeast Pacific ridge, or most of the Indian Ocean ridge, do not develop a crestal rift. The cymatogenic ridge does not develop tensional fracture of stress at deeper lithospheric levels where deformation is mainly by vertical laminar displacement. The axis of the ridge is thus viewed not only shallowly under tension but fundamentally as a megashear (p. 188), which view accords with the measurements of Hast (1969) (p. 188), and the observations of Icelandic geologists (p. 188). The conjugate, transverse system of shears provides the cross-fractures so typical of most of the submarine ridges.

At this point the second interpretation dispenses with transform faults, required by the first interpretation, and regards the *transform* portion of the cross-faulting as due to the flank-arching effect in opposite directions on either side of the central axis.

The pattern of geomagnetic polarity reversals (Fig. 22) is then seen not as a 'sea-floor spreading' phenomenon but as a series of geomagnetic shadows which are not 'frozen' into rocks but which move laterally away from the ridge freely through the rocks to either side (Chapter 6).

Euler's theorem applies to the distribution of phenomena much as it does in plate tectonics, and the two interpretations come together here. 'Sea-floor spreading' and 'global expansion' are then seen to be the same thing (Fig. 31).

Of the two interpretations sketched above, the first has received constant support and publicity, and is regarded as orthodox; the second offers new points of view, and may be tomorrow's orthodoxy. In Chapters 3 to 9 pursuing different lines of thought, we therefore devote more space to the latter interpretation.

Chapter 3

Sea-floor cymatogens

All at sea

The cymatogens described in Chapter 2 are mostly on land, where they can be inspected and studied at leisure. But it is the sea-floor cymatogens that have held attention of recent years. Although until recently these could be explored only by remote-sensing methods, they could be examined more freely than the cymatogens that were obscured by a continental cover.

Isacks, Oliver and Sykes (1968) defined the thin oceanic lithosphere as overlying an asthenosphere which in turn covers the main region of mantle, or mesosphere, and this definition has met with general acceptance. It emphasizes the thinness of the oceanic lithosphere, and the presence beneath it of an upper mantle layer sensitive to changes of stress and capable of considerable mobility in response to any changes of stress that may occur due to either outside or internal influences. The mesosphere they regarded as relatively inert. 'A key point of this model is that the pattern of flow in the asthenosphere may largely be controlled by the configurations and motions of the surface plates (p. 77) of lithosphere and not by a geometrical fit of convection cells of simple shape into an idealized model of the earth.' This is so, of course, where there is de-gassing of magma and sinking of heavier magmatic residues (p. 78).

Exploration of the ocean floors during the years since World War II has revealed the existence there of several large linear rises. Some of these are comparable with the cymatogens already familiar on the continents, even possessing a rift valley with much shallow seismic activity along parts of the crest. They also exhibit a narrow zone of augmented heat flow and much associated volcanism. Moreover, the current seismic and volcanic activity indicates that the present submarine ridges with their uparchings and rifts are of recent development and are probably coeval with similar cymatogenic activities upon the continents (p. 49). These developed greatly in the early or mid-Miocene (e.g. Pacific Ridge), and were mightily rejuvenated during the late Pliocene and Quaternary. In this view, tectonic events on the continents and beneath the oceans are similar in type, semicontemporaneous and both derive from subjacent activity within the upper mantle.

General properties of the principal sea-floor cymatogens

Under the oceans the Earth's crust is generally very thin, but it thickens remarkably beneath the mid-oceanic ridges.

The general form, dimensions, and structure of the three principal sea-floor cymatogens (Atlantic–Arctic, Indian and East Pacific) are:

(a) From a depth of 3000 to 4000 metres below sea level broad swells arise with a maximum width of about 700 km. Late Cenozoic/Quaternary arching, indicated by gravity surveys in the vicinity of volcanic islands, seems to follow the isostatic crustal model of Gunn (1949) with an elastic crust overlying a fluid substratum (asthenosphere), e.g. mid-Atlantic Ridge above a depth of 3500 metres.

In simple form the principal suboceanic ridges, like cymatogens on the lands, involve no necessary lateral stresses except towards the crest of the arch where, the arc being longer than the chord (Fig. 6) some lateral tension may occur in the rigid upper crust causing a rift valley to crack open and shallow, modern seismic and volcanic activities to appear. The appearance of the central rift is due to rigidity of the crust plus the increased length of the arc over the chord as expressed in the risen arch. No such tension is induced in the fluid mantle below, where, indeed, lateral compression is usual (p. 37).

Seen thus, the structure of a sea-floor cymatogen may be assessed as a broad, arched zone followed by the trend of the sea-floor ridge. In the upper, elastic crust a broadly shattered structure is evident with numerous vertical potential shear planes (sometimes in bundles) parallel with each other and with the general trend of the sea-floor ridge. This trend is probably derived from plastic laminar displacements (vertical) in the asthenosphere and mantle below (p. 14). During deformation of the elastic crust there is much shifting of displacement from one bundle of shear planes to another (Fig. 14).

Of course, not all the large ocean ridges and rises are cymatogens. The geology of Madagascar which stands on the northern end of a long ridge, for instance, identifies it as a sliver off the nearby African continent so that it is a minor fragment left by the Mesozoic Gondwanaland disruption. South Georgia and Kerguelen are similar. New Zealand too (with the Campbell submarine plateau) began as an eastern fringe to Australia but after separating in the late Mesozoic, and allowing the Tasman Sea basin to develop, New Zealand became involved with the Tonga–Kermadec structures where it was reactivated. It is currently highly seismic, and has enhanced heat flow and active volcanoes.

So, although we shall refer to the subject again later (p. 145) it is well to realize from the outset that submarine ridges are not infrequently bodies of mixed development.

Smooth gentle rises, with or without a rift zone (e.g. parts of the East Pacific Ridge; and the Carlsberg Ridge of the Arabian Sea) are probably no older than Pleistocene. On the other hand, prominent sea-floor ridges which contain rocks of undoubted continental derivation (as do parts of the Atlantic and Indian Ocean ridges) must have originated at the Cretaceous dismemberment of Gondwanaland.

Several of the submarine ridges are aligned along megashears produced by the reaction of a rigid crust to deformation in the asthenosphere below. Some are related to horizontal crustal movements of the plate tectonics type (p. 113). Partaking of both states are regions like New Zealand and the west coast of North America, a former margin of Laurasia now being undermined by the new cymatogen of the East Pacific–Juan de Fuca Ridge (Atwater, 1970) (p. 158).

(b) Both flanks of the mid-Atlantic Ridge down to a depth of 3000 metres (Beloussov, 1970) are longitudinally zoned by terraces with a relief of up to 1000 metres. The steps are separated by steep ledges, and such stepped topography is not unusual on the flanks of all the major suboceanic ridges. The condition would appear to be faulted.

(c) The surface of the Atlantic Ridge is typically rough from the crest out to the lower flanks. Much bare rock bottom is exposed, and 'total sediment accumulation is very small, averaging 100–200 m.' (Ewing et al., 1964). Often the sediments are thicker in pockets, many of which have almost level surfaces. 'The sediments are unstratified and remarkably transparent acoustically. Certain areas, particularly on the lower flanks of the ridge, contain distorted sedimentary bodies that apparently indicate post-depositional tectonic activity.' van Andd and Bowin (1965) seem to be in favour of gravitational sliding of crustal masses off both flanks of the ridge.

(d) The lower flanks of each ridge are usually buried in sediment which is largely pelagic. This sediment may be conformable and outward dipping due to later tectonic rise of the ridge axis, or it may overlap on to the ridge, in which case younger strata successively make the base of the covering sediments as the central zone is approached (Fig. 34). At one time this 'younging' of the basal sediments up the flank of the ridge was widely quoted as evidence of sea-floor spreading; but simple overlap on to an intermittently rising cymatogenic ridge is a more natural and simple explanation.

(e) The deeper composition and structure of the suboceanic ridges (p. 29) is determined on seismic data and this was early done for the central part of the North Atlantic Ridge by Ewing (Fig. 13). Attention should be directed to the relatively large body of 'levitated mantle' with seismic velocity 7.3–7.5 km/s, which he describes. Few earthquakes have been reported away from the ridge crest.

(f) The mid-ocean ridges stand in approximate isostatic equilibrium and are therefore stable features of the bathymetry.

(g) Sea-floor cymatogens in general have been shown by Hast (1969) to be oriented so that they strike at an angle of 45° to the vector of maximum compressive stress within the crust. *If this be so (and Hast's measurements seem conclusive) one of the trends of maximum shear must lie along the trend of the ridge and the other transversely, thus coinciding with the set of conjugate shears.*

Contrary to widespread belief (sea-floor spreading), the rift zone in the upper, rigid lithosphere appears to be a shallow phenomenon related only to the vertical movements on the arc–chord relation of a rising arch (Figs 7, 11, 12, 14). The main body of the cymatogenic arch is truly vertical and shears take place largely in the vertical sense. This mechanism (Fig. 14) is perhaps better related to the intermittent appearance of the 'rift zone' especially along the Indian and Pacific Ridges.

(h) It has been established that ocean floor cymatogens migrate, and in changing position they collide: (i) with one another; (ii) with sea-floor trenches (Chapter 4); or (iii) with continents (p. 158). In some cases this rules out a convective origin for both phenomena in one system.

The principal submarine ridges hold in common with cymatogens on the lands all the features (a) to (f) and the ultimate proof of their identity is that, in certain instances, continental cymatogenic features appear to be continuous with suboceanic ridges, and where this is so the generalizations (a) to (e) above appear to pass regularly from one to the other. In other words a single controlling mechanism is present in the upper mantle beneath continent and ocean floor alike. This mechanism is expressed zonally and near vertically through the crust (p. 19).

However, one feature the continental crust and oceanic crust do *not* share. At irregular intervals the mid-ocean rift zones are interrupted by faults which *appear* to have laterally offset both the rift-valley morphology and the magnetic anomaly pattern (Fig. 42). Studies by Heezen and Bunce (1964) in the equatorial Atlantic 'revealed that the fracture zones (which continue as far as the Guinea coast) grow longer at the crest of the mid-oceanic Ridge and that their extensions across the adjacent basins are inactive.' Cross-fractures of this sort affect both the Indian and Pacific Oceanic Ridges also; but no such features appear in rifted continental crusts, with the possible exception of the Ethiopian rift.

Seismic evidence from the mid-ocean ridges

We now examine another line of research — earthquake records — and shall find again that different basic assumptions are possible, leading once more to different conclusions. Sykes took the view that the direction of maximum crustal tension is at 90° in the horizontal to the ridge direction; he accepted the 'transform' faults of Wilson with reversed movement and duly arrived at 'the hypothesis of ocean floor growth at the crest of the mid-ocean ridge'. And Isacks, Oliver & Sykes (1969) opined: 'For earthquakes located on such major transform faults as the oceanic fracture zones, the San Andreas fault and the Queen Charlotte Islands fault, one of the slip vectors is very nearly parallel to the transform fault on which the earthquake was located . . . On the contrary, earthquakes along the ridge crest but not on fracture zones do not contain a large strike slip component but are characterized by a predominance of normal faulting.'

But two years later Hast (1969) showed that the direction of maximum compression is at 45° to the trend of the crestal rift, and this leads inevitably to the conclusion that the trend of the ridge and the rift is a broad zone of megashear. And in Hast's 20,000 measurements of crustal compression there was not a single exception to this regimen! *Not one* registered a state of tension in the lower crust as is assumed under the hypothesis of sea-floor spreading by convective currents. Surely Hast's measurements confirm that a general state of compression exists within a sphere ruled by centripetal gravity.

If the main direction of the ridge is a megashear, then the numerous cross-faults are shears also (Fig. 14) with one remarkable property: earthquakes originating on these shears are concentrated on the sectors between the two ends of successive 'rift zone' segments, and are few elsewhere along the length of these conjugate shears. This suggests that displacements on those active sectors of the conjugate shears are mainly vertical. Displacements upon the sheared 'rift zone' are also vertical and normal. . . . The main structures of the ridge are thus ideal for intrusion by underlying basic magmas; but there is no measurable sign of horizontal sea-floor spreading (Fig. 14). Some other factor must be responsible for that sort of phenomenon, and that we shall seek in later chapters.

The Red Sea Basin

At first glance, the African and Arabian massifs seem formerly to have been united, and the basin between them appears to have been formed by simple sea-floor spreading between the opposite sides. Was this so?

Darracott, Fairhead, Girdler and Hall (1972) have expressed a general opinion: 'To a first order, the East African Rift System may be considered to be a consequence of movement along the boundaries of three plates *viz.* the Arabia, Nubia and Somalia plates.' However, the Gulf of Aden (Laughton, 1966; Laughton *et al.*, 1970), while showing the oceanic nature of the crust within the Gulf, has magnetic anomalies that are very recent (anomalies 1–5). Moreover, two lines flown over the Red Sea at latitude 18°N show that in anomalies over the width of the axial trough there is some suggestion that these anomalies and the oceanic-type floor cover a wider area than the trough itself.

The third component is the Eastern Rift down through Somalia and Kenya, which shows no separation of the sides with introduction of oceanic-type volcanics.

Seismic velocities of 5.9 km/s (continental rocks) for eight refraction profiles

(Drake and Girdler, 1964) indicate that the opposed shores of the Red Sea could never have been in contact; nor is the 500 m isobath any more reliable for fitting. Nevertheless, Arabia displays much Pliocene to Recent basaltic eruptivity.

The Afro–Arabian dome elevates Cenozoic landsurfaces and is a youthful feature of the scene. The Red Sea Basin is slashed across the dome in rift-valley style. So, with its proximity to the rifts of North Africa, and evident continuation in the Jordan rift the Red Sea was early claimed as the incipient opening of a new sea. But opinion has steadily moved away from this viewpoint, largely because of growing doubts whether Africa and Arabia were formerly one land mass. Thus Coleman (1974): 'The early formation of the Red Sea is somewhat obscure and is not clearly a result of sea-floor spreading.' He notes the absence of faults upon the northeast side parallel to the Red Sea shore, and records the scarp 50–160 km inland as erosional. Finally, while the Gulf of Suez is controlled by normal faulting, the Gulf of Aqaba is a result of wrench faulting.

Cochran (1973), however, has returned to the sea-floor spreading hypothesis.

In summary; refitting of the opposite sides seems not very easy (the Danakil Horst gets in the way; and the 500 m isobath is not much better. Matching of ancient geological formations on the two sides is less detailed than might be expected if the basin was a simple rift. Indeed, there seems to be no demonstration that, in antiquity the two sides, as we see them now, were parts of a single land mass.

Arabia, from the distribution of its later Palaeozoic and Mesozoic rocks then occupied a marginal place in Gondwana; but as Fig. 2 shows, its primary position may have been to the side of the present proximity to Egypt and Sudan. If this were so, evidence of the shear faulting between Africa and Arabia would be important, and this structure is known, not only in the western Red Sea floor but also in the Gulf of Aqaba and the Dead Sea rift (Quennell, 1958).

J. A. Whiteman (1971), writing of late Precambrian times in this region noted: 'Support for the hypothesis made by previous workers on rift systems of a northward translation movement, and an anti-clockwise rotation of the Arabian Peninsula relative to Africa is supported in our data in the western part of the shield' etc.; and I. Gass (1976) too: 'In lower Tertiary times, after a long period of tectonic stability and magnetic quiescence extending from late Precambrian times, this region became the site of intense magnetic and tectonic activity that has continued virtually without interruption to the present day.' 'The oceanic crust in the (southern) Red Sea and Gulf of Aden formed during the lateral separation of sialic blocks in late Tertiary times is of tholeiitic character.'

From the sedimentary evidence: in the coastal plains of the Red Sea there is considerable Miocene evaporite like that in the Mediterranean (p. 171), but no marine equivalents of the Indian Ocean. So clearly the Red Sea area was part of the dried-up Mediterranean during the late Miocene, and the rift was not in existence before the Pliocene. Even in the southern Red Sea, where Deep-Sea Drilling Project (DSDP) Site 228 was drilled on the western side of the central rift to see whether the sequence of sediments was the same as at Sites 225 and 227 east of the central rift the scientists reported 'it was'. 'The finding of a lower Pliocene and probably upper Miocene sequence at this site presents considerable difficulties to supporters of continuous sea-floor spreading of the Red Sea because the site overlies a magnetic anomaly supposedly 2.2 million years old.' Only 'at the beginning of the Pliocene do marine oozes and marginal elastics begin to be deposited on the Miocene evaporite sequence', and only at that late date does a truly marine Red Sea begin.

So the Red Sea Basin, associated with the African Rift Valley, has often been described as modern, newly opening sea floor, displaying a median rift and with evidence of sea-floor spreading. It is almost too obvious to mention.

However, of special interest is the report by Lowell, Genik, Nelson, and Tucker (1975), based upon an abundance of significant oil-company data. The authors begin with the regional upwarping of continental lithosphere in the southern Red Sea area: 'with concomitant volcanism, erosion and sedimentation of continental rather than marine character'. All this was followed in the report by 'divergent convective flow in the asthenosphere' which was 'marked by pronounced subsidence of horsts and grabens with accumulation of a younger (Quaternary) salt in the Danakil Depression.'

At the same time, in a second rift west of the axial trough, where a Pliocene clastic sequence was developed, 'oceanic-type basalts were extruded in the south-west corner of the Red Sea.'

The authors summarized: 'Our studies have shown that the break-up of continental lithosphere was by normal faulting which dictates that sialic blocks must be present seaward of the present coastline (Fig. 9 of original report): thus the basement from coast to coast cannot be oceanic crust.'

Consequently, 'the present coastlines of the Red Sea cannot be used to make a pre-rift restoration and neither can a fit be made with isobaths.'

The central and northern Red Sea may be taken as floored from side to side by continental material, not oceanic crust, and except for the southwest corner in modern times the Red Sea Basin should not be regarded by geologists as an incipient ocean developing by separation of the sides accompanied by sea-floor spreading from the axial rift trough.

The world pattern of major suboceanic ridges

The bathymetric charts of Heezen and Tharp, copied and recopied in charts and atlases everywhere, have made the major topographic forms of the ocean bed almost as familiar to us as the topographies of the continents. And prominent upon them are the great submarine ridges of the mid-Atlantic–Arctic, mid-Indian and Antarctic–East Pacific Oceans.

The mid-Atlantic Ridge begins in the Arctic at the mouth of the Lena River which itself flows in a faulted trough. Thence it extends straight across the floor of the Arctic Basin towards the east coast of Greenland. An accompanying seismic belt causes it to be regarded as an active cymatogen. At 80°N, having passed the pole, it continues southward until it reaches Jan Mayen where it shifts from west to east across the southern front of that island.

In these latitudes Talwani and Eldholm (1977) have reviewed the evolution of the Norwegian–Greenland Sea which began 60–63 million years ago. Up to 30 million years ago the Labrador Sea was also opening and the motion of Greenland was northwest from Eurasia. At that time only the Norwegian Basin opened up, and Greenland slipped by the Barents Sea and Svalbaard transcurrently. Afterwards, Greenland moved almost west from Norway, and the more northern basin opened up.

The deep North Atlantic has here narrowed considerably, due to broad continental shelves on both sides. The Barents sea floor was above sea level during most of the Cenozoic, although seismic velocities suggest Mesozoic and Palaeozoic sediments beneath.

Iceland stands upon the ridge at about the latitude of the polar circle. The geology and geophysics of this volcanic island have been the subject of thorough examination by the Icelanders themselves who find (p. 77) that it does not conform as well as might be expected with results from, and theories of, the Reykjanes sector of ridge immediately to the south.

Now in its typical form, the mid-Atlantic Ridge has a seismic belt 150 km wide (though south of Iceland the Reykjanes Ridge shows no crestal rift at all!) But the ridge itself arches across a width of 1500 to 2000 km. As always, the cymatogenic arch is far greater than the (presumed) tensional zone. The broad flanks are markedly stepped, uniformly upon both sides of the ridge are three steps descending from the crest to the abyssal plains of the ocean basins.

From the latitude of Iceland the Atlantic Ridge keeps a sinuous median course southwards between North America and western Europe until at 35°N, by the Azores submarine plateau, the ridge records a cross-zone of shattered rocks (the Gibraltar–Azores Fracture Zone) that apparently marks the junction with the southern part of the mid-Atlantic Ridge derived on the separation of South America and Africa.

From now on, the number and importance of cross-fractures increases (all in an east–west orientation): Oceanographer, Kane, Vema, Romanche, Ascension and Rio Grande, some of which continue into either continent: some into both.

The greatest of these cross-fractures is the Falkland–Agulhas Fracture Zone which transects the South Atlantic between latitudes 55°S and 30°S. This, as Rabinowitz and la Brecque (1979) have shown, is the mighty fracture zone on which the separation of Africa and South America was achieved

(Fig. 38). It finally trends north-northeast in the Indian Ocean towards the Moçambique Channel.

Through all its length the base of the Atlantic Ridge floors about one-third the full width of the ocean.

Southward the same ridge continues beyond the Falkland–Agulhas Fracture Zone (which is apparently the older of the two features) to Bouvet Island, where it dies away at latitude 60°S. A great many maps depict it continuing under the name of 'Southwest Indian Ridge'; but in the words of Laughton, Sclater and McKenzie (1972) 'The southwest branch is in fact a complex set of *en echelon* northwest to southeast spreading centres and northeast to southwest fracture zones.' The sea-floor topography between the main mid-Indian Ridge, a normal cymatogen, and Africa is replete with elongated basins and intervening ridges, and displays many islands, some continental others volcanic and oceanic in character.

Equatorially, the median rift sometimes dies away. Of four traverses, at 10°, 13°, 16°, 19°N, only that at 13°N shows a well developed rift.

Early studies of the Atlantic Ridge focused largely on the remarkable parallelism of the coastal margins to east and west with the ridge trend itself, and the deduction was made that all these features (in both North and South Atlantic) were closely related, and had originated in Jura-Cretaceous time at the disruption of Laurasia and Gondwana. Because the separation presumably still continues, the assumption was made that the Atlantic has been widening on either side at an average rate of 2 cm per year. The idea of sea-floor spreading immediately became popular. But, as we shall often point out in this text, averages mean very little in geology. Most geological processes take place briefly when definite physico-chemical boundaries are passed. They are often local in application, and provoke an irreversible change which alters the problem entirely.

The Indian Ocean Basin is likewise traversed from northwest to southeast by a single, even straighter, suboceanic ridge. It too has a broad topographic uplift with high heat flow, shallow seismicity and sporadic rift valley formation along a narrow zone towards the crest. Beneath the broad arch lies a body of levitated mantle with a seismic velocity of 7.2–7.9 km/s, which is the evident source of all the tectonic activity at higher crustal levels. It is a typical cymatogen, but more diversified than is the mid-Atlantic Ridge.

The mid-Indian Ridge appears at the Gulf of Aden, as the Carlsberg Ridge; but this is low and flat, and associated geological data suggest that this topography is no older than Pleistocene and that the previous continuation of the mid-Indian Ridge lay along the Chagos–Maldive line. Indeed, late Proterozoic and Palaeozoic igneous and metamorphic samples recovered from near the crest of the Carlsberg Ridge suggest that the ridge is here really a fragment of the neighbouring continents rather than a normal sea-floor ridge structure.

In these latitudes is the transverse Owen Fracture Zone which links the long transcurrent faults of western Pakistan to the mid-Indian Ridge and the Somali Abyssal Plain which continues the zone of fracturing towards the Moçambique Channel and perhaps to junction with the Falklands–Agulhas Fracture Zone. Not all these features are of the same geological age, but they certainly seem to indicate a line of separation off East Africa (Fig. 2); of great importance in the topographic evolution of this region. This line is also marked by deep-seated intrusions of ultrabasic rocks along at least the Owen Fracture Zone (Hamlyn and Bonatti, 1980).

Curving from north–south to west–east, the Indian Ocean Ridge separates a western basin replete with islands and submarine plateaux and ridges from an eastern region almost devoid of lands and constituting, until Australia invaded it (p. 7), part of the Proto-Pacific realm. So the Indian Ocean Ridge differs materially from the mid-Atlantic Ridge with its bilateral symmetry extending to the margins of the boundary continents. Sea-floor spreading (Norton and Sclater, 1979) is complicated by fracture zones in the western Indian Ocean, and indeed different lengths of the Indian Ridge appear to be of significantly different age as though the whole had been aggregated of different pieces at different times ranging from Permian to Pleistocene. Ultimately the ridge passes equidistantly between Australia and Antarctica to the Macquarie Ridge where it terminates. This latter is not a cymatogenic ridge but may be a sliver of Gondwanaland, or a shear like the Ninety East Ridge, extending 4000 km due south from the Bay of Bengal. Ninety East is not cymatogenic. It apparently originated by transcurrent faulting. (p. 154).

The Antarctic–East Pacific Ridge is by far the largest. It is very wide but flat and low; and it reaches from Macquarie Ridge in the Antarctic sector in a sweeping curve to Easter Island and thence northward to California and Oregon where it disappears beneath the continental shelf of North America. This course, unlike the Atlantic Ridge, does not equally bisect the ocean in which it occurs; it is exceedingly eccentric.

All measurements of age along the ridge rate it as relatively young (10 Ma). Heat flow is high at the crest; but for great lengths (including the nearly 9000 km stretch from Macquarie to Easter Island) the ridge lacks a crestal rift valley, although earthquakes are equally common.

At depths of 4000 metres or more the ridge sometimes spreads out into broad plateaux. The Albatross Plateau around Easter Island yields seismic data which suggest that the crust there may incorporate lithologic material of truly continental type. This may result from extreme magmatic differentiation, for the island itself has obsidian derived on extreme differentiation from basalt (Bandy, 1937). Throughout, the ridge, affords evidence of a core (at depth) of levitated mantle rock, sufficing for classification of the structure as a cymatogen.

Four subsidiary submarine ridges (northwest Chile, Sala-y-Gomez, Nasca and Carnegie-Cocos) link the East Pacific Ridge with South America. All these bear the credentials of youthful (Miocene or later) cymatogeny.

Crustal section and structure through a cymatogen

The structure of a cymatogen, whether upon land or at sea, is really a section through the Earth's crust, and it is probed by seismic refraction methods. A typical section at sea was early provided by the mid-Atlantic Ridge (Ewing and Ewing, 1959) whose data showed a normal subcrust (7.6–7.8 km/s) beneath the basins alongside, changing in the axial zone to a 4.5–5.5 km/s layer overlying a 7.3 km/s layer of considerable thickness. On the flanks of the ridge the crust is an uplifted oceanic crust with subsided ocean basins beyond. Surprising was the size of the core-mass of levitated material (Fig. 13), which occupied a large proportion of the volume below the axial zone of the ridge.

Later, Menard (1961) pointed out that the East Pacific Rise had been produced by a process acting mainly in the mantle, and investigators came to consider that the material of seismic velocity 6.8–7.3 km/s present below mid-oceanic ridges was anomalous upper mantle (normally 8.1 km/s). Later a linear relation was shown between the depth of oceanic water at any point and the depth to the top of the mantle, and the range of density (6.8–7.8 km/s) of the anomalously levitated mantle showed what a great range of volatiles could be occluded in subcrustal magmas. The density of mantle material could be reduced even below that of the overlying rigid crust. Levitated and mobile, the magma was ready to move under static load through any available fissure in the crust. The rest of the tale was just diapir tectonics.

The velocity of such rises (as Cloos showed for salt diapirs) increases until, about half-way up to the surface, the diapir spreads laterally below the roof and in a domed form. The action ends theoretically with the light, mobile layer transferred through, and lying above, the denser former roof (Berner *et al.*, 1972).

Since the Ewing's, several researchers have elaborated models of the mid-ocean ridge structure based on petrological, seismic and latterly borehole data without achieving unanimity. One of the latest, and most reliable, is that by Dewey and Kidd (1977).

Land-based cymatogens are similar at depth. But superficially, erosion cuts into the uplifted parts and reveals an intermediate layer of crystalline tectonic gneiss (Fig. 6). This gneiss is quartzo-feldspathic (microcline), generally of coarse texture, with the bands of light and dark minerals repeated every 5 mm or so, and all disposed nearly vertically. *Such mineral bands are the result of vertical laminar movement in continental-type crust, and provide the tectonic mechanism for the uplift apparent by arching at the surface.* The temperature at which they form, in response to pressures, has been estimated at 450–600 °C and they correspond to the material recorded seismically as 4.5–5.5 km/s. These gneisses are non-magmatic; they are devoid of pegmatites, pneumatolytic ore deposits, and commonly also of granitization effects. *They are purely tectonic in origin,* caused by steeply-inclined laminar movements in a confined continental-type crust associated with production of fresh minerals of low temperature formation. Amphibolites are often associated.

This characteristic facies, found in the rejuvenated monoclines of Natal, Angola, Rio de Janeiro and Mt Kosciusko, Australia, demonstrates the mechanics of continental cymatogeny at intermediate to shallow crustal depths, corresponding to surface warpings involving thousands of metres of vertical displacement over zone widths of a thousand or more kilometres (Fig. 6). The number of active laminae may be very large, and consequently the differential movement upon each lamina need be only very small. With a laminar thickness of perhaps 5 mm, which is fairly typical of these gneisses (about 200,000 laminae per kilometre) very great displacement can be achieved over the width of a cymatogenic arch. Thus for a surface tilt of one degree the differential movement *per lamina* need be less than one two-hundredth of a centimetre. But the cumulative surface rise over 100 km would exceed 1650 metres.

Cymatogeny upon the lands affects both shield areas and orogenic areas alike, which implies an independence from crustal composition; and if our correlation with the sub-oceanic ridge structures is correct, then a subcrustal origin for the energies involved is clearly implied.

Most important is the observation that the crustal deformation is vertical. No suggestion of horizontal movement of the crust is found, and therefore no connection with continental drift is implied. Indeed, the more the geology of Cenozoic time is examined, the more reason appears for regarding the horizontal drift function as dominant mainly in late Mesozoic time, and the present configuration of the continents as having been then designed with but little alteration during the Cenozoic era.

The deep root of material with seismic velocity 7.0–7.9 km/s which has been determined beneath cymatogens is not known to be presently exposed anywhere at the Earth's surface. Several lines of investigation, however, have prompted a general acceptance by geologists that it should be regarded as 'anomalous mantle' material: that is, as ultrabasic material of velocity 8.1 km/s type which has been charged abundantly with primitive volatiles.

The main source of cymatogenic activity is therefore to be sought in the asthenosphere or upper mantle, the rock material differentiating as it rose from olivine tholeiite to alkali basalt.

The range of seismic velocities (7.0–7.9 km/s) indicates clearly that the material is not simple, but that it is compounded in varying proportions of two or more constituents, one of which is about as dense as olivine, the other of which (though compressed into the very body of the olivine) is very light. This light constituent is capable of conferring considerable mobility in the mass, especially where it is locally concentrated within either the crust or the upper mantle.

Where this mobile 'anomalous mantle' has invaded the crust up to regions below 500 °C temperature and of greatly reduced pressure it alters extensively (according to Hess, 1960) to serpentine, with an increase in volume of 13%. Intrusive serpentine belts (p. 118) have been mapped invading regions of tectonic gneiss belonging to ancient cymatogens. (This transformation to serpentine is viewed less favourably nowadays than it used to be.)

The roots of young cymatogens, of late Cenozoic and Quaternary age, show also in negative gravity anomalies which sometimes follow the surface form of the cymatogen quite faithfully, e.g. eastern Australia, western United States, eastern Canada (King, 1962/7).

The nature and role of volatiles included in deep crustal and subcrustal rocks

By far the commonest volatile substance in rock masses is water vapour (steam). Others are silicon dioxide, carbon dioxide, sulphur dioxide, ferric oxide, ammonia and ammonium salts, oxygen and nitrogen. In cymatogenic belts much of the water seems to have been highly silicic. These two conclusions derive from:

(a) analyses of gases contained in the structural planes of crystals in ultrabasic rocks;
(b) the abundance of quartz in granites and gneisses of the continental platforms;
(c) the abundance of water vapour in volcanic exhalations and the frequently associated silicic waters;
(d) the abundance of amygdales (e.g. agate) in basic effusives; and
(e) the cumulative abundance of water in the hydrosphere and atmosphere, both of which have accreted progressively from volcanic eruption through geologic ages.

Locally the volatiles carried concentrations of ore minerals, but the proportions of these are seldom significant.

The geochemical effects of H_2O and CO_2 on magma generation in the crust and mantle have been discussed by Wyllie (1977).

Volatile constituents within the Earth exercise a number of most important roles:

(a) They lower the densities (and seismic velocities) of the rock materials wherein they reside and may create reversals of density in the column of rock materials. So materials that are strongly charged with volatiles will tend to rise either in the upper mantle or within the crust. Such behaviour is probably the ultimate cause of cymatogenic arches and domes with their attendant fracturing of the crust towards the Earth's surface (Fig. 10).
(b) They reduce viscosity and may thereby induce rock flow within both the mantle and the lower crust.
(c) As volcanic effusion of primitive steams into both the atmosphere and

hydrosphere is strongly associated with cymatogens, and as heat flow from the Earth's interior is likewise enhanced in cymatogens the volatiles are understood to be the principal heat transferrers within the body of the Earth.

(d) The presence of volatiles in rock melts and magmas lowers the temperatures of fusion for rock-forming minerals, thereby enabling them to melt at markedly lower temperatures than in dry fusion.

In all these roles the effect is to increase the potential mobility in rock masses, and there can be little question that intraterrestrial volatiles exercise an important control over subterranean tectonic activity, of which surface and shallow crustal deformation is a consequence.

The investigations and conclusions of Einarsson (1949) at the eruption of Mt Hekla in Iceland are relevant here. Geologically, Iceland is a pile of volcanic ejecta standing squarely upon the crest of the mid-Atlantic cymatogen in latitude 66°N. Information concerning its formation, growth, and activity should therefore be specially instructive regarding the nature and behaviour of Earth materials involved in: (a) volcanic eruption; and (b) cymatogenesis.

Every geologist is familiar with the oft-repeated textbook phrase 'the lava [or magma] cooled and solidified [or crystallized].' Lavas and magmas, however, do not crystallize because their temperature drops, they do so because they lose their volatile constituents: a basaltic lava issuing at a temperature in excess of 1000 °C may flow with turbulence, lose its volatiles rapidly, and freeze solid before its temperature falls to 900 °C. The same lava flowing smoothly by laminar flow can still move slowly at 600 °C.

For these and other data we are indebted to Einarsson, whose researches at the eruptions of Hekla (1947 and 1948) and Surtsey (1963–5) shed fresh light upon the processes operating when magmas rise toward, and are emitted at, the earth's surface. Summarized, the progress of eruption at Hekla was as follows.

After an initial outrush of bombs and pumice due to access of meteoric waters, the eruption of Hekla in 1947–8 produced a vesicular basaltic lava whereof Einarsson (1949, 1951) recorded temperatures between 1130 °C and 1150 °C at many places. Within the crater temperatures of 1000 °C were measured, the higher temperatures outside, flames and explosions with block showers being due to burning of gases mixed with the atmosphere. At the lava fronts the maximum temperature measured was 930 °C. A lava viscosity of 10^7 poises at about this temperature permitted either fracture or flow according to the rapidity of application of the forces, some broke off as blocks (glowing inside), some formed extrusion tongues. This gave rise to much differential flow. Fluidity control is difficult to define. 'It includes a factor other than temperature and even viscosity, for a lava, once slowed, never regains its former rate of flow.'

The density of lava varied greatly, from 1 or even less on emission as spongy lava to 2 for dense lava flowing plastically at a distance of 6 km from the crater. The difference was due to loss of volatiles, chiefly water vapour. On analysis the Hekla lavas were found to be very low (0.35%) in water, even the fresh spongy lava at the point of emission. Yet it was all very fluid. By personal tests Einarsson determined viscosities of:

5×10^5 poises on the most fluid lava;
1.3×10^6 poises on the porous plastic lava front;
1.5×10^7 poises on the most viscous, dense lava.

In such lavas turbulence was developed at Hekla. These compare with 10^4 poises for the most fluid Hawaiian lavas.

For the actual solidification of lava and its relation to gas loss, Einarsson noted that primary lava flooding at the points of extrusion was spongy, with densities as low as 0.6 or 0.8. In the light, foam-like rock, gas-filled vesicles were small, with thin walls. Flowing lava remained of this type. Consolidated blocks some distance down the flow, however, were dense (2.0) and lacking in vesicles, the gas bubbles having coalesced and left the lava. At an inter-mediate stage the lava possessed 'relatively thick walls separating largely intercommunicating cavities'. This is an important stage in the solidification of the lava, marked by a shearing movement by which the gas bubbles coalesce, thus enabling steam to pass at pressure through the many conduits which replace the original small vesicles. Some of the conduits are lined with small points produced by tearing. These points, near the end of flows, also show melting which can only have been caused by the passage of very hot gases (1260–1200 °C). Such additional heat is available from exothermic reactions, or perhaps from the latent heat of crystallization.

Consolidation of the lava is thus not a simple freezing of a normal fluid. 'Other factors than cooling seemed to be of greater importance' (Einarsson). In neither the light spongy lava, nor the dense blocks was the crystallinity related to rate of cooling. The lava indeed showed a strong reluctance to crystallize. When it did crystallize, which it probably did suddenly, it did so in relation to shearing and loss of the gas vesicles. This was the final act in the process of de-gassing of the mantle.

When it occurs, crystallization is rapid and viscosity increases markedly. With a latent heat of 90 cal/g a rise of temperature due to crystallization could be by 360 °C. A rise observed by Einarsson was 200 °C. The same principle presumably applies to crystallization at depth within the Earth, with the production of effects recognizable towards the surface of the Earth.

The production of ash also follows from the shearing process within the lava, when small teeth were produced upon the disjunctive surfaces. Such ash is, of course, basic; but the introduction of much silica in the magmatic waters (of which much evidence exists in the amygdales of basalts) could well during the same process provide an acid ash of ignimbrite type. Such ash, often called rhyolite ash, is however truly an acidified basic ash. In the

Lebombo of southeast Africa so-called rhyolites not only contained 8% iron ores corresponding with associated basalts, but parts of the acid materials themselves were proved to be metasomatized basalts. The waters active in this process are of course highly heated juvenile waters forming part of the magma column itself. The actual explosion of ash at the surface is attributed to contact with meteoric water which is heated only to between 200 and 300 °c. The proportion of acid materials is often much too high to permit of explanation on differentiation of a basaltic parent alone. In Iceland 20% of the total postglacial tephra are rhyolite.

Some evidence which suggests that primitive volatiles may occasionally be very siliceous is listed:

(a) The abundant siliceous waters of hot springs and geysers.
(b) Highly explosive eruptions of great magnitude producing huge quantities of acid tuff but little or no lava seem to be largely aqueous, e.g. Valley of Ten Thousand Smokes in Alaska, the New Zealand ignimbrites (in part) and the Lebombo acid tuffs.
(c) The enormous quantity of agate and related silica minerals in plateau basalts, e.g. Drakensberg of South Africa, São Bento basalts of Brazil.
(d) Kimberlite magma (together with a metamorphosed and silicified variety called metakimberlite (Bardet, 1973) is probably emplaced explosively at low temperatures ($\sim$ 600 °C) compared with most other magmas (Mitchell, 1973). Abundant xenoliths of primitive-type peridotite record an earlier stage in its history.
(e) The highly siliceous nature of many sulphide-bearing bodies seemingly derived from depth by pneumatolitic action.
(f) Patches of theoleiitic, not olivenic, basalt even in the ocean basins where contamination by assimilation of continental-type crust does not appear possible.
(g) Pods of ultrasilicified materials (with tough granophyric texture) in bodies of the earth's crust where it is not usually so acidified.

Now and again one has to remind oneself that not only is silica soluble in hot water, but water is also soluble in silicate melts. Such pods could represent 'damp' patches in a silicate melt under considerable pressure.

Carbon dioxide too exercises different roles, and under the great pressures typical of the mantle it reacts with peridotite. Cavities filled with liquid carbon dioxide are common in peridotite xenoliths from alkali-rich basalts. Green (1974) is of the opinion that these are not a result of melting but of 'out-gassing' and that they provide part of the driving energy for plate tectonics (Chapter 7).

Repetitive emissions of different lava types from a single vent are quite usual. Amid the Eocene basalts of Iceland tholeiites, olivine basalts and porphyritic basalts have poured out with change of magma composition more than 50 times during the whole 5000 metres of the volcanic pile. No rhythm is apparent, although each type was erupted widely for a long period before the composition of the magma changed and a fresh group was emitted.

More striking are the instances where the magma contrasts are extreme, as where basalt and rhyolite repeatedly share the same vent. Inclusions show that the acid phase succeeds the basic. Where andesites are present they are not intermediate but hybrids, and several lines of evidence indicate not mixing but coexistence of basic and acid magma types. Though mutual solution among silicate melts is generally assumed, non-mixing between magmas may be as real as non-mixing between turbid and clear, saline and non-saline waters.

Einarsson's penetrating observations from Iceland, standing upon the mid-Atlantic Ridge far out in the ocean, are of great importance in the study of fundamental volcanic and tectonic processes within the Earth. And Einarsson seems here to have identified the little 'energy demons' of Earth activity (p. 16). They are the volatile products residing in the tiny vesicles of primary magma within the mantle of the Earth. Their escape under turbulent *seething*, has been called 'de-gassing' of the mantle (Vinogradov, 1964).

As the mantle is present everywhere beneath the crust (both oceanic and continental) of a spheroidal Earth, tectonism (active or latent) must exist everywhere, and its potential within the crust and at the surface will largely depend upon the nature (especially thickness) of that crust from place to place. All weaknesses are likely to be exploited, all differentials probed. This suggests that formerly-used weaknesses are likely to be used again, and some localities have remarkable records of repeated tectonism and volcanicity.

Glass is not a result of rapid cooling, nor is much water necessary for its formation. It is the natural end state of porous lava that has not undergone stirring. I had myself an experience of this phenomenon. Knowing of glassy limburgite lavas in the Northern Lebombo, I collected some examples hoping that by analysis the primitive gases of mantle peridotite might be ascertained. The samples were examined in Russia by the courtesy of Professor A. P. Vinogradov; but the report (1968) said that the vesicles were too small for collection of the gases to be made. In other words, the magma was primitive porous type frozen to glass and with the gases still in vesicles that were too minute to supply adequate gas samples. Glasses show that the first mineral to crystallize is usually magnetite.

The lessons learned from the solidifying of Icelandic lavas, especially the relatively low importance of cooling, and the very high importance of volatiles and of shearing at certain critical viscosities, the early crystallization of magnetite and the several types of glasses all have importance to be assessed in the behaviour of (a) plutonic and subcrustal magmas, and (b) in the production of crustal structures, especially those classed as of vertical and zonary distribution, i.e. in cymatogens like the mid-Atlantic Ridge.

The Deception Island eruptions of 1968–9

Despite having twice previously burnt the nails out of the soles of my boots by the heat in active volcanic craters, I have been impressed more by the power exhibited in the forceful outrush of volcanic steams and associated products. These outrushes begin with explosion at the opening up of the vent, they continue through all the active phases of eruption, and long after the eruption of ash and lava have ceased the force with which the gases continue to be emitted seems not to diminish, and when the geyser stage is reached their emission may still be awesome.

It was therefore with the greatest interest that I witnessed the eruption of Deception Island in West Antarctica during January–February 1969. The eruption began with an explosive outburst from beneath the icecap. Cubes of glacier ice 1–2 metres in diameter were scattered far and wide over a radius of more than a kilometre. Then within minutes a series of lahars was launched and came tumbling down the mountain sides. Masses of ice, lava and scoria (old and new) slid like glaciers of rubble. One of these destroyed the main hut of the British base, bursting through it and filling the place with debris. Then the scoria rained down. Lastly, torrents of water burst over the landscape, and scored channels through the debris down to the shore of Port Foster. By good fortune the five members of staff had evacuated the hut immediately before the eruption started, and when they returned after half an hour they found that their hut had been destroyed.

My first observations were made two days after this outburst. I visited the orifice (Fig. 16(a)), determined the sequence of events, and spoke with the men whose hut had been destroyed.

Returning after a fortnight, I found further great changes. In place of the eruptive centre at first blown out was a chain of chasms reaching eight kilometres from the original centre to the Chilean base which had been destroyed by eruption the previous year. The whole of this system of chasms, together with several parallel fractures, was discharging enormous volumes of steam many hundreds of metres into the air (Fig. 16(b)). The general strike of the active chasms was nearly parallel to the straight east coast (also fractured) of the island. Although little glowing lava could be found, quantities of bombs and ash had been discharged and thickly covered the part of the island east of Port Foster. Deposits of sulphur and ferric oxide marked the sites of fumaroles.

Despite the noxious gases, descent was made within the chasm to the very bottom (near sea level) where the opportunity was taken to verify every point that Trausti Einarsson had made in Iceland, insofar as this could be done at Deception Island.

(a)

(b)

Figure 16(a) Volcanic explosion pit through the ice cap, Deception Island, Antarctica, January–February 1969. The crater walls visible through the overcast and gently-snowing weather are about 50 m high; the bottom is in volcanic rocks well below this depth. The surrounding country was littered by blocks of glacier ice (like those in the foreground) which preceded the scoria (Lester King).

(b) Two weeks later. Clouds of volcanic steam rising hundreds of metres into the air from a fissure 8 km long, February 1969. Incandescent lava was visible at the bottom of the fissure. RRS *Shackleton* in the foreground (Lester King)

The FAMOUS Expedition

Far better information came from the French–American mid-Ocean Undersea Study group operating with deep-water submersibles near the Azores. In ocean depths of 1500 m the crestal rift of the mid-Atlantic Ridge was found to be of about the same size as rift valleys known upon the lands, with steep sides and abundant faulting parallel to the trend of the ridge. From certain of the fissures issued hot lava (1200 °C) which was quickly congealed into 'pillow' structures by sea water, the temperature of which was near freezing. Under great pressure the lava was injected about one metre at an impulse.

The centre line of the rift was followed by a steep, narrow trough, the rocky floor of which revealed a shattered network of innumerable intersecting fissures such as are found when tunnelling along a shear zone. This was the most active part of the rift. Lavas at the centre line were the youngest. The submersibles photographed it in depths of 2600 m of sea water.

The observations revealed almost exactly what had been envisaged by ship operators for the rift zone of a mid-ocean ridge. The phenomena agree as well with Earth expansion as with convection currents and sea-floor spreading.

Chapter 4

Zones of subduction

Subduction

In July 1939 at the eruption of Nyamlagira in eastern Zaire, I witnessed a demonstration of true magmatic *subduction*. The pool of basanite lava was about 35 metres across: it heaved and tossed and flowed turbulently, but heavily. Rocks thrown on to the molten lava (temperature: 1160 °C) did not sink, they merely skidded across its surface as upon ice. New red- to white-hot lava rose freely (usually about the margins of the pool), and congealed in a matter of minutes or seconds to a black irridescent skin thinly covering the surface of the pool. Intermittently this skin was withdrawn across the surface of the pool to an area of subduction where it wrinkled, thickened and was *sucked down* into the pool again (Fig. 17). This was true subduction, by gravity, of lava which had lost its volatiles to the atmosphere, become heavy and sunk into the rising, gas-charged magma.

The process continued, with variations, for months. The lava was very fluid and poured away from the vent through underground tunnels and down the sides of the volcano 22 km to Lake Kivu where its heat was quenched amid great clouds of steam.

It is doubtful whether the word 'subduction' as used by modern geophysicists refers to any comparable phenomenon or process of being 'sucked under' or 'drawn down' into the mantle by normal gravity. Readers will doubtless differ in their interpretation of the mechanics of 'subduction' and crustal engulfment. Subduction nowadays is usually not seen at all, but is inferred from the occurrence of a wide range of phenomena in the sea floor together signifying youthful zones of subsidence which are often marked by deep submarine trenches. The earliest direct evidence there of subductive subsidence came with the measurement of trenchward tilting of flat-topped Capricorn guyot outside Tonga Trench. Offshore reflection profiling then revealed that the flanks of trenches commonly display a horst and graben structure with sediment layers conformable to one another and correlatable to the intervening horsts, but all with an overall dip into the trench. Shipborne refraction profiling over the Peru–Chile Trench indicated further that

(a)

(b)

Figure 17(a)   Volcanic lava pool on Nyamlagira, Belgian Congo, July 1939. Note how the free-flowing hot lava issuing from the margin of the pool pours over the already-formed skin of solid lava (dense and de-gassified) on the nearer side of the pool (Lester King).

(b) Lava crust on the left sinking steeply (? subducting) into freshly flowing lava from the right. The contact between the two phases is almost a straight line (Lester King)

56

the sea-floor crust of oceanic basalt *descends from the inner side of the trench* beneath *the continental margin*. Reliable data from the Aleutian Trench,* moreover, 'indicate that ocean crust now beneath the axis . . . began subsiding in late Pleistocene time, and that sediment filled the trench rapidly . . . Also, the drill-cores consist of competent, highly deformed Pleistocene sediment', which recalls the 8000 metres of deformed lower Pleistocene sediment in the Ventura Valley of California. Sediments are sometimes contorted and piled against the inner wall of the trench, as though by thrust faulting, and the southern flank of Alaska has been described as a 'typical subduction zone' where oceanic crust is at present being destroyed under the American continent. We should note the recent date of this violent activity. Conditions were long quiescent during the middle Cenozoic, and cores from the Bering and Kamchatka areas show a steady accumulation of turbidites from Miocene continental erosion. Some trenches have received great quantities of turbidites from the end.

Piling and distortion of sediments along the landward wall of trenches is by no means unusual and has sometimes been ascribed to *scraping off* of pelagic sediments under the action of sea-floor spreading. Other dredged samples suggest that the lower, continental-side walls may contain igneous rocks of the deep oceanic crust, or magmatic cumulates, instead of metamorphosed sediments; and all lower, inner walls are sites of strong thrusting and shattering.

But nowhere is there evidence that oceanic sediments have disappeared in quantity beneath the continents during the late Cenozoic or Quaternary time.

Further properties of deep-sea trenches

Whereever they are found, existing trenches are currently active and afford topographic, geomorphic, and tectonic evidence of their geological youth. They transgress late Cenozoic marine sediments, and although older trenches are known which have been filled, many modern trenches have only one or two kilometre thickness of flat-lying strata on the floor. The sides, if symmetrical, are very steep and go down to between 7000 and 8000 metres beneath the surface of the ocean.

Heat-flow values are small in the trenches, in contrast with the rifted crests of sea-floor cymatogens; but andesitic volcanism is often associated (Fig. 18).

The majority of trenches have large negative gravity anomalies (often 150–250 milligals) which can be accounted for largely by lack of mass within the trenches themselves. This is characteristic of their youth and there is no evidence of isostatic collapse either by heaving up of the floor or crushing-in of the sides, even where the anomaly is recorded directly beneath the position of the trench. Where island arcs are associated with curved trenches

* DSDP Leg 18, 1971.

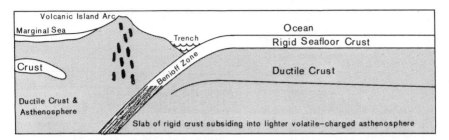

Figure 18 A subducted slab of rigid crust submerging under gravity into the levitated mantle or asthenosphere. Both the lower crust (ductile) and the asthenosphere accommodate the subsidence by vertical laminar flow (not shown). Volcanism (often andesitic) develops above the descending slab and may appear as an arc of volcanic islands

the anomaly may be to one side, beneath an adjacent, recently-uplifted island arc, e.g. Guam.

Like cymatogens, subduction zones have apparently a large subjacent mass of anomalously light mantle rock. Petrologists have suggested rock changes to explain this mass: e.g. basification (Beloussov, 1964), deserpentinization (Hess, 1955) and 'eclogite–basalt phase changes' (Ringwood, 1966); but what is needed is primarily a change of physical state, for example a mantle peridotite enriched in siliceous volatile constituents (p. 45), so that it is less dense than the overlying, cold oceanic crust, which thereby sags down into it in the form of a trapdoor subsidence near the oceanic margin. The Cayman Trench, which serves as a boundary between two Earth-plates (Cuba and Hispaniola) has on its north flank a terrific escarpment which goes right through the sea-floor sediments and exhibits towards the base extruded masses of upper mantle rocks. This natural Mohole displays the ultrabasic rocks at 3658 m — the greatest depth of the trench reaches 4724 m.

The Kurile Islands have instead: (a) a distinct alkaline-earth enrichment from basalt to rhyodacite, and (b) free silica caused by low alkali content. The lavas have been derived from basalt magma which has assimilated silica on the way up, as is proved by a geophysical sialic root.

The above facts show a causal similarity with the cymatogens of the previous chapter. Both seem to have the same fundamental agency — large-scale intrusion by active volatile-charged mantle rock.

Both operate by the same mechanism, vertical laminar crustal and subcrustal displacement, but of opposite sign: the cymatogen rises to a sea-floor ridge and maintains its cover of densified crust; the subductogen crust goes down to make a deep trench and develops an inclined slab of crust which has broken away from the main body of crust on one side (Figs 18 and 19).

How can the same agency produce opposite effects? Perhaps by operating in different structural environments. Cymatogens are symmetrical structures with a tendency to intermittent rise so that the crust is freely arched above the rising mass of levitated mantle. Only later is the arch broken at its crest,

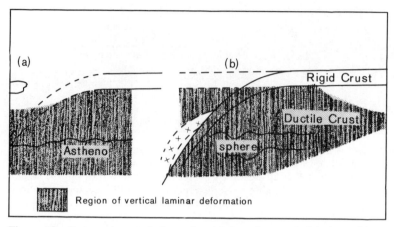

Figure 19   Progressive evolution of a subducted crustal slab from (a) to (b) does not necessitate any lateral displacement of the rigid crust. Only the *zone of change* (wherein the change is brought about by *vertical* laminar flow) moves laterally. Note the zone, indicated by crosses, wherein levitated mantle material is obducted above the subsiding slab

where a rift valley is formed by stretching of the rigid crustal layer (p. 19). Subductogens are always asymmetrical, with a slab of crust sinking like a trapdoor to complete submergence in the asthenosphere. This structure forms where the rigid oceanic crust is early broken, and the levitated mantle exploits the fracture to rise through the heavier, rigid overlying crust and is free to obtrude above it. Now the relative reversal of gravity relations, with already de-gassed, rigid crust overlying fresh magma which is mobilized and levitated with a high content of primitive volatiles, becomes crucial. Gravity takes over and the crustal edge sinks as a slab through the lighter, rising volatile-charged mantle or asthenosphere which thereupon obtrudes above the slab. This is the process I observed in the lava pool at Nyamlagira in 1939.

Seemingly a favourite point for such a breakthrough is on the oceanic side of an old continental–oceanic margin, and the breakthrough commonly leads ultimately to severe orogenic deformation at the surface.

The distribution of deep-sea trenches

A simple trench east–west in Drake Passage, between Cape Horn and the Antarctic Peninsula, shows a long smooth descent upon the north side leading down to a flat floor of sedimentary infill 20 km wide. On the south side a steep scarp climbs rapidly upward to the shelf of King George Island. Only when the eastward turn of the cordilleras of Tierra del Fuego and Graham-land is envisaged, and the occurrence of serpentine on tiny Gibbs Island is remembered, does the trench assume special significance as a possible shear zone between two continents.

There are several such major fractures on the floors of the oceans, e.g.

Cayman Trench of the Caribbean, where there has been much metamorphism and injection by upper mantle rock types (Perfit, 1976); but they are not regarded as true subduction zones.

The straight Philippine Trench, deeper than 8500 metres for more than 800 km of its 1250 km length, though thought by Rutland (1968) to be a strike-slip fault, has the typical trapdoor structure of a subductogen, as has the straight Tonga Trench which also yielded fresh dunite and peridotite at depths between 9100 and 9400 m on the western wall, which is where the Benioff zone would crop out. Xenoliths in the basaltic rocks of Hawaii also include cumulate phases of dunite, lherzolite, wehrlite, harzburgite, pyroxenite, and garnet peridotite from the upper mantle.

Roedder (1971) has generalized: 'Olivine in ultramafic xenoliths in alkali basaltic lavas from all over the world contain inclusions of liquid CO_2 under high pressure. It is concluded that the presumably basaltic melts from which the olivine crystallized were *saturated* with carbon dioxide and a separate dense supercritical CO_2 phase was present as immiscible globules in them. Pressure estimates place the depth of origin of many of the inclusions studied in the range 8–16 kilometres, so the crystals cannot possibly have originated by mere crystal settling in the throat of a volcano.' Even single crystals of such olivine are much larger than the average crystal size of the basalt host, to which they may appear quite foreign, e.g. at Cape Crozier, Ross Island.

In the western Indian Ocean the greatest known depth called Vema Trench (6400 m) crosses the Carlsberg Ridge near latitude 9°S.

It is somewhat over 100 km long, on a bearing 55° true and is very narrow. The trench probably follows a fracture zone with right lateral displacement. It resembles the Romanche Trench which displaces the Atlantic mid-ocean ridge. Neither of these, however, possesses the qualifications which apply to the main ocean-floor trenches, viz. marginal to continental crust and possessing a Benioff earthquake zone with deep-focus earthquakes.

Northwest of Australia, in the Wharton Basin, are several short, linear deeps. Curiously, some of these contain turbidites dated 80 Ma (Upper Cretaceous in DSDP borehole 212). Others have only pelagic blanket over the topography; and both types have at least 10 metres of red clay at the top of the sedimentary column. Tectonism is currently quiescent. An anomalous history is evident here; but even the pattern of magnetic reversals is described as difficult to interpret and provides no solution.

We may perhaps compare these Wharton Deeps with other Jurassic basins of eastern Asia, and consider that the Wharton examples became separated during the Miocene when Australia–New Guinea drifted northward into collision with southeast Asia (Fig. 3(d)).

The west coast of South and Central America is bordered for thousands of kilometres by deep-sea trenches topographically younger than the giant ridge and plateau of the southeastern Pacific on which, indeed, some of the trenches occur. But these trenches are not straightforward strike-slip faults, they are associated with an eastward-dipping subductive structure which passes beneath the continent, to depths of 700 km. The Palaeozoic history of both the Americas tells, indeed, of former trenches and island arc structures that were overridden, and their sediments were incorporated into the cordilleran structures by late Mesozoic westward continental drift. North America also overrode, in part, the site of the younger East Pacific Ridge (p. 158). The modern trenches of Central and South America are tectonically active, and beneath them the ocean crust is said to increase in thickness by 50% and the Moho is flexed down, indicating perhaps that the fundamental cause for the trench-sinking there lies at mantle depth.

The northwestern Pacific is remarkable for its strongly curved island arc–trench systems reaching from Alaska to, and including, Indonesia. Each arc has an individual asymmetric and subductive structure, with the 'sinking slab' of oceanic crust dipping towards the Asiatic mainland from which the arc is separated by marginal seas more often shallow than deep, sometimes floored with oceanic basalt.

Along the Kurile arc, where the distance between the trench and the volcanic arc is 240 km, is a prominent marginal swell (the Hokkaido Rise) 200–400 km wide and several hundred metres high. This big rise between oceanic floor and the trench (where subduction begins) suggests plain vertical tectonics in recent time. Off Japan the same Hokkaido Rise bounds a trench in which the young sediments are cut by many postdepositional faults, predominantly normal, showing that the deepening of the trench is recent. The Indonesian Trench is also young. Opposite Timor it subsided during the late Pliocene.

While the modern trenches are related to modern tectonics, the marginal seas may be older and relate to Palaeozoic–Mesozoic westward drift of Asia which was then the trailing edge of Laurasia (Fig. 3); but the formation of the marginal basins was much later. The sea of Japan, for instance, is thought to have opened in the early Miocene and been completed in the late Cenozoic. An exception is the South China Sea. The eastern edge of this basin has a crust comparable with that of the Pacific Ocean (except that it is only half as thick) (Ludwig, 1970). The basin is locked off by the Philippine Islands, a mobile belt with an abundance of basic and ultrabasic rocks and little in the way of continental rocks. So any suggestion that the South China Sea basin was subsided is untenable. These structures continue to Taiwan.

The greatest congregation of trenches and allied features, and the widest variety of types is found between the Philippines and Tonga. There the Marianas Trench attains the deepest sounding in the oceans (11,033 metres of the Challenger Deep). In scale and setting it resembles the deepest part of Horizon Deep in the Tonga Trench, the ocean's second deepest point; while the Cape Johnson Deep in the Philippine Trench is the third; and the New Hebrides Trench also descends below 9000 metres to a gently sedimented floor.

There is a riot of strange features present in Micronesia and Melanesia where the Sulu Sea is a strongly subsided marginal basin with oceanic crust, but in the shallower northwest it exhibits a seismic structure typical of submarine ridges. The Colville Ridge in the east, between Tonga and Fiji, gives every evidence of being a former arc rifted and torn apart, with subsequent creation of a new (Pliocene) ocean floor, and there is a new arc upon the western side (Karig, 1970). The arcs of New Britain and the Solomon Islands are turned inside out and thus have their trenches on the landward side.

The authors of the preliminary report DSDP Leg 30, Wellington to Solomon Islands (*Geotimes*, September 1973, pp. 18–21), suggested from their data that 'arc migration does not explain the origin of the marginal basins of the southwest Pacific'. For the area Kermadec–Tonga–Fiji–New Hebrides they were of the opinion that 'intermittent divergence between the Australian and Pacific plates (Chapter 7) due to changes in rates of motion, poles of rotation and resultant vectors, satisfies the requirements with subduction of only small amounts of crust'. They outline the development in four phases: 'Phase I more than 80 m.y.a. During this period Australia, New Zealand, Antarctica existed as a single land mass . . . Phase II 80–56 m.y.a. . . . began with the onset of northward movement of the Pacific plate past the still stationary Australia/New Zealand landmass with resultant separation of New Zealand from Australia and the opening of the Tasman Sea.' The phase ended with separation of Australia from Antarctica and the beginning of its trek northward. 'Phase III 55–10 m.y.a. Northward motion of Australia and Pacific plates. No subduction, but the interior South Fiji, New Hebrides and Coral Sea basins opened up' and accepted Eocene and later sediment. 'Phase IV 10 m.y.a.–present. Change in rotational axis of Pacific plate. Shearing developed at the western margin of the Pacific leading to subduction along the New Hebrides–Solomon line.' This makes a cohesive story for which there is much evidence.

Deposition in the Coral Sea is of interest (site 210). Basin crust formed in the early Eocene. Oozes and cherts were formed during the Oligocene (see p. 163) as though supplies of terrigenous waste from Australia were meagre. 'Turbidite distals (from revived erosion on the continent) began with mid-Miocene' and 'built to the level of the sea-floor by early Pliocene.' There are 90 metres of Pleistocene turbidites. Sites 288 and 289 north of Solomon Islands also show a major Eocene to mid-Oligocene hiatus and several other widespread breaks of succession in a region far from land.

The two Atlantic trenches bounding island arcs are:

(a) The Puerto Rico Trench rimming the 80 million year old basins of the Caribbean and joining (apparently since the Eocene) the severed ends of the North and South American cordillera where these have been turned east into the Atlantic. By contrast, opposite the southern Antillean islands, from Guadalupe to Grenada, the trench fades away, and between the islands and the abyssal plain of the Atlantic is a broad region of deformed Eocene–Miocene sediments that beneath Barbados reaches a depth exceeding 20 km. On gravity and seismic refraction data Westbrook, Griffith, and

Barker suggest that the subsided crustal slab may lie beneath this outer zone. Certainly low-alkali basalts, apparently obducted above the slab, are known on these Lesser Antillean islands. They concluded, however, that 'attempts at correlating circum-Caribbean geology with plate motions (Chapter 7) have met with only partial success'.

(b) The desolate South Sandwich arc and trench lie far out in the South Atlantic where, like the Antilles, they link the severed ends of the Andean (Tierra del Fuego) structures which are seen in South Georgia (Dalziel, 1970) with the extension of the Antarctic Peninsula eastward through South Orkneys. Beautiful Z-folds in South Georgia are directed to the north, South Orcadian structures are oppositely directed to the south (p. 165).

The youthfulness of the still active, barren volcanic islands and the freshness of the trench walls suggest that this is the youngest of all the trenches. This belief was confirmed: (a) by the thinness of the bottom sediments recorded by USNS *Eltanin* in contrast to the smooth thick sediment carpet of the Argentine Basin to the north and the Weddell Basin to the south; and (b) by magnetic polarity reversals 0–7. Within the arc, the eastern sea floor is believed to be of late Cenozoic age.

All the major trenches except the Romanche and Vema Trenches (which are ancillary shears crossing the mid-Atlantic and mid-Indian Ridges respectively) are: (a) clearly related to the boundary between continental crust and oceanic crust; and (b) found only along those parts of the world's coasts that were originally part of the Palaeozoic coasts of Gondwana and Laurasia. Most of them therefore appear upon the margins of the present-day Pacific. The simplest relationship is shown by the Peru–Chile Trench, parallel to the cordillera of South America. More complex are the looping island arcs with marginal sea basins of eastern Asia. There are Banda arc of Indonesia directly overlies the zone of negative anomalies testifying to recent tectonics. But the presence in this zone of Permian and pre-Permian rocks in association with Cenozoic sediments which are overthrust towards the modern trench in alpinotype nappes, shows that the core formations are related to those of Malaysia. The modern trenches and their anomalies are moulded upon pre-existing structures — not the reverse.

Structural relationships are similar in Japan, the Philippines and other east Asiatic arcs.

The disruptive oceans — Atlantic and Indian — even where the structure of their coastlands is clearly monoclinal towards the ocean (p. 205), as for both east and west Africa, have no trenches. There are no deep-earthquake belts at the continental margins, and there is nothing to indicate relative motion other than tilting (p. 194) between the sea floor and the bordering continent. Under the plate tectonics theory (p. 112), the continents are there regarded as part of the same Earth-plate as the adjacent ocean floor.

However, modern deep-sea trenches are often paralleled upon the landward side by fossil (Palaeozoic and Mesozoic) trenches that have been filled by terrigenous waste (turbidites) which often thicken towards the source.

Frequently the bedding shows tectonic disturbance, and the Ewing brothers cite them as proof of discontinuous sea-floor evolution. An example in the east Pacific lies between the Juan de Fuca Ridge and the mainland. Its highly stratified turbidites now dip *east*, showing that the present Juan de Fuca Ridge rose into existence later, presumably by vertical uplift. Ewing *et al.* (1968) estimate that it rose about 3 million or less years ago, which is too young to fit with Vine's magnetic polarity chronology for the region. They continue with a paragraph: 'It is worth noting at this point that the marginal trench west of Tierra del Fuego is also filled and has essentially the same pattern of sedimentation. A further point of similarity in these two sections of trench is the absence of associated intermediate or deep focus earthquakes. According to the sea-floor spreading hypothesis, as the crust is apparently not under-running the continents in these areas, it must be assumed that the ridges are moving away from the continents as Wilson (1965) has suggested.'

So both the concept of sea-floor spreading and the geomagnetic chronology here create, rather than solve problems.

Benioff zones

Earthquake shocks associated with trenches have a distinctive pattern. Unlike the shallow shocks associated with cymatogens they derive upon an asymmetrical structure and originate at shallow (0–100 km), intermediate (100–300 km), and deep (300–700 km) foci; and like the structure, which flexes down to pass beneath an adjacent continent, they appear to be generated upon a plane (a Benioff zone) which is similarly inclined at angles from 15° to 90° to the horizontal. Seismic data also confirm the observation from dredge hauls and piston cores that thrusts and folds affect crustal rocks of the lower, inner slope of the relevant trenches.

The focal mechanism of intermediate and deep-focus earthquakes is not explosive, but of the type caused by slippage on a fault. The slip vectors show that the quake motion is not generated by movement of the slab through the asthenosphere but is internal, within the slab, showing that it retains the crustal property of deforming by fracture far down into the asthenosphere. Ultimately it breaks up by internal stress, like an iceberg.

Sometimes I recall an iceberg off Adelaide Island in 1962. Situated 30–40 metres off the port quarter of RRS *Shackleton* on a still, sunny morning was an old, blue iceberg rising about 10 metres out of the sea. A rustling sound coming from the 'berg alerted me. As I watched, small pieces of ice began to pour down the side of the 'berg, many of them breaking up still further on the way. Presently the iceberg was surrounded by a ring of ice fragments, each only a few centimetres in diameter, dancing on the surface of the quiet sea. Clearly the ice of that 'berg was everywhere breaking up under internal strain produced by the release of one-time confining pressures. Occasionally the iceberg tilted and changed its balance in the water; but all the time the process of disintegration continued. After 15 or 20 minutes the iceberg was no more, and the sea all around was covered with a brash of small fragments of ice. An iceberg had died, piecemeal, in a quarter of an hour. Can a body of crustal rock, strained to the point where it becomes traversed by a rhomboidal or parallelogram system of fractures, not disintegrate similarly when submerged in the asthenosphere?

The slab is not, as the Geophysicist's Credo affirms, 'made mantle'. Its former change from lava to crust (involving loss of volatiles to the atmosphere, and loss of the original porous ('spongy') structure, are not to be regained merely by melting. A permanent loss of energy to the tectonic system had been incurred by loss of volatiles and destruction of original porous structure (Einarsson) (p. 48).

The dip of a subsiding slab generally ranges from 30° to 40° (Isu Bonin arc) to nearly vertical (Marianas arc). Associated island arcs are indeed rooted not in the crust but in the mantle. Oliver *et al.* (1974) have considered that 'the downgoing plate is perhaps fractured or divided along near-vertical surfaces oriented approximately normal to the arc', and in extreme cases 'the transmission of seismic waves in the region indicates that the lower part of the slab is detached', e.g. New Hebrides. 'The radiation patterns from shallow-focus shocks near the trench show that they are caused by underthrusting of one crustal plate by another.' And any change of direction by the trench is faithfully followed by the deeper structures and earthquake shocks.

The Benioff zones go deep, some down to 700 km, with the seismic layer near the top of the zone. Thus the S-wave from deep-seated earthquake foci has been described as 'following the sunken crustal slab up to the surface.' So deep-focus earthquakes originating from west of Japan are felt heaviest in the east of Japan. Comparable effects are described from the Tonga Trench, where the Benioff seismic zone is said (J. Oliver) to dip right through the asthenosphere to the mesosphere.

The Tonga–Kermadec ridge and trench system is 2700 km long and very straight. It is spoken of as a boundary zone between a continental plate on the west and the Pacific Basin on the east. Superficially the trench is a graben containing little sediment and the ridge is controlled by horst and graben structures. Katz (1974) opines that these features are indicative of tensile stress across the continental margin. 'This' he writes, 'is difficult to reconcile with the commonly held view that Pacific-type orogenic margins are areas of collision and consumption of major crustal elements. There was certainly no evidence of compression, as has been found also along the margin of Chile, in which country the tectonic history throughout the Cenozoic indicates a state of tension and breakdown.' This is consistent with a thinning of the crust underneath the Chile Trench (Worzel, 1965).

But the Tonga–Kermadec features have become a classic for the clear development of a subducted slab structure, based upon a large amount of consistent geophysical data, which can be seismically followed from the sea bed continuously down to a depth of 650 km within the Earth.

On the Tonga Islands the oldest sediments are Eocene, usually overlain by

Miocene which was deformed before being covered by Pliocene–Pleistocene limestones, over 200 metres thick. Even these are tilted on Vava'u. The Tonga Ridge is bordered to the west by a fault zone with 1500–2000 metres of downthrow and the same beds appear to continue under the Lau Basin which, however, certainly has a new oceanic floor of basalt.

Numerous vertical up and down movements typify the region; and local upwellings of mantle material occur. Southward, the Kermadec Trench fades away in the abyssal plain at 3500 metres, but in New Zealand Hatherton (1970, 1971) pointed out that 'southward the seismic subduction zone goes ashore in North Island, and nearly bisects the island'. At the south end of North Island the seismic zone swings eastward towards the Pacific. Hatherton (1971) concluded that 'if sea-floor spreading [p. 75] is taking place, the eastern part of North Island has "floated" like a raft and has somehow escaped subduction, despite the fact that seismic data indicate that it is coupled to the mantle.'

Brown *et al.* (1968) showed that the eastern part of North Island is under-lain by Permian (?) and Mesozoic geosynclinal sediments which pass westwards across the zone of subduction into rocks of the same ages and facies. Similar situations 'appear to involve Roti, Timor, Tanimber and Kai Islands in the Indonesia arc; the Vogelkop of West Irian; Halmahera and possibly parts of other arc systems'. The global tectonics model with inclined subduction plane seems not always to fit the facts of geology (see also Beloussov, 1970).

Rock types associated with subductogens

Petrologists have devoted much attention to the types of rock developed in different parts of subductogens (e.g. Ringwood, 1974), and have sought an explanation therefrom for the characteristic tectonism. But the problem of crustal subsidence is here a physical, not a chemical one and its solution will ultimately have to be stated in physical terms which are not incompatible with the petrological facts. Data are drawn not only from modern trench and arc structures, but also by structural analysis of former subductogens (orog-enic zones) which have been denuded after uplift, so that the deeper structure of those former subductogens are now exposed.

Little seems to happen on the oceanic side except that blue schist forms *beneath* the slab where it bends downward. The composition of these glauco-phane schists is consonant with their formation from alkali-rich oceanic crust.

But where the slab has been subducted to greater depths and is drawn below continental crust it seems to acquire greater activity at depths about 100 km (a depth where large numbers of intermediate earthquake foci are produced). Ten times as much seismic energy is displayed in the island arcs and similar structures as over all the rest of the Earth. In particular, greater activity of deep-seated volatiles may be deduced from the fact that this zone underlies the surface volcanic arc. Can this enhanced activity therefore be expressed in the terms already derived from Einarsson's volcanic studies in Iceland (p. 48)? There would appear to have been already a change from the primitive porous lava of the oceanic type, which has ruptured the thin walls of the original vesicles and allowed the release of volatiles at some time in the past. Crushing of the lower end of the slab by the great Benioff thrust fault perhaps causes rapid final crystallization (p. 48), with expulsion of the more mobile residues. So begins the rise, from the subducted slab, of fresh magma resources and gases. The last of the basalt magma and volatiles tend to escape (diapirically) upwards through the heterogeneous continental crust where calcium and alumina may be added to the rising magma which emerges superficially as calc-alkaline andesite. Indeed, most andesite volcanoes occur where crust is being destroyed, that is, over descending slabs of crust (Fig. 18).

Parts of the country rock are metamorphosed appropriately in an aqueous environment to calcareous green schists and amphibolites (still with water vapour) and locally minor granophyric and plutonic intrusions are formed secondarily by the same vapours.

Heatflow temperatures measured in the volcanic zone are naturally high, e.g. behind the island arcs of the western Pacific. As we noted in the sea-floor cymatogens both the rise of the agencies and the results which they cause are *vertical*. In subductogens they mark the destruction of the subsiding slab directly below. And again the phenomenon of dilatancy, with rapid and reversed changes from solid to liquid phase, is probably active, e.g. in the formation of granophyres.

A considerable variety of mantle-type rocks is found. Serpentinized peridotite and 'talc rock' have been recovered from depths exceeding 2000 metres within the trenches; while from the deepest part of the inshore wall of Tonga Trench come 'fresh harzburgites, some dunite, serpentinized lherzolite, serpentine lenses and variolitic low-alkali basalts'. A comparable collection came also from the deep inner wall of Cayman Trough in the Caribbean, and the Owen fracture zone of the Arabian Sea (p. 43). Samples from the lower part are often slickensided and strained; and seismic refraction studies recorded compressional wave velocities of 5.2–6.1 km/s. 'Neither the topographic, rock sampling nor refraction profiling studies suggest that pelagic sediments were plastered against the deep near-shore walls here in the Tonga Trench. Neither did the dredging yield rocks derived from metamorphosed or melted deep sediments.' In other words, there was no evidence of laterally-directed sea-floor spreading (p. 75).

Properties of subductogens

The salient properties of subductogens are:

(a) They are places where asymmetrical, downward tectonic movements occur.

(b) The attendant sea-floor feature is a trench of unusual depth.

(c) A volcanic island arc sometimes stands between the trench and the mainland.

(d) Most subductogens border long-established continental margins, that is, they occur at boundaries (p. 112) between oceanic-type crust and continental crust. Even the Tonga, Marianas, and South Sandwich Trenches, far out in the present oceans, have (according to seismic data) relics of continental crust upon the western side.

(e) Earthquakes are endemic with subductogens. The foci are distributed along a plane dipping landwards (a Benioff zone). The shallowest foci are situated towards the base of the inner wall of the trench where evidence of rock breaking and distortion is also found. Intermediate foci (100–150 km depth) are common beneath the volcanic arc, and deep-focus earthquakes (300–700 km) occur sometimes beneath the neighbouring continent.

(f) Beneath both contrasted types of crust (continental and oceanic) the Earth's mantle is continuous. The upper part (asthenosphere) is fluid and is charged with volatile constituents, which reduce its density, and increase its mobility, sometimes to the extent that, as Press (1969) found from the examination of seismic records, the suboceanic crust is sometimes denser than the mantle rocks of the asthenosphere beneath.

The difference of density was explained petrochemically by Press — he proposed that half the dense lithospheric crust was eclogite — a denser form of basalt. We prefer to regard it as basalt from which the volatile constituents had escaped during eruption whereupon the basalt solidified. So densified, there need be no chemical change to eclogite.

In either case the result would be a one-way, irreversible change, gravitationally unstable in the vertical sense. Isostatic readjustment was then achieved by sinking of a slab of basaltic edge (high density) into the mobile asthenosphere at the crustal discontinuity between continent and ocean. Such sinking and depression from the geoid shape inevitably introduces the arc–chord relationship once more (p. 35); but this time the relationship is reversed and the new state produced is not an extension of rigid crust at the surface, but horizontal compression of the sinking crust at depth — manifested by underthrusting of one crustal plate by another. Normally the continental side rises, the oceanic side goes underneath.

Fig. 20 shows how the existing subduction zones are, almost all of them, related to the outlines of Laurasia and Gondwana.

(g) An asymmetric, inclined slab at the edge of the suboceanic crust is the structure characteristic of all subductogens.

(h) An inclined slab subsiding parallel with itself may appear at the surface to be migrating outwards at the trench and volcanic arc (Fig. 19). This appearance has caused authors of a previous generation to lay undue emphasis upon 'the growth and migration of island arcs', and is still quoted. Kimura has summarized for Japan: 'An impressive aspect of ancient Japanese

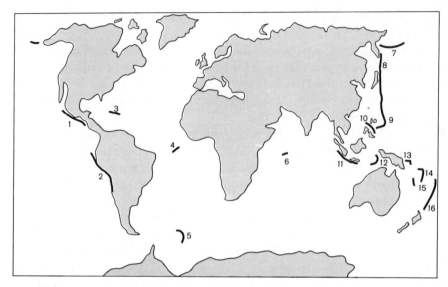

Figure 20   Sites of the world's great subduction trenches most of which are related to the outlines of Laurasia and Gondwana: 1, Middle America; 2, Chile–Peru; 3, Puerto Rico; 4, Romanche; 5, South Sandwich; 6, Vema; 7, Aleutian; 8, Japan–Kurile; 9, Marianas; 10, Philippine; 11, Indonesia; 12, Banda; 13, New Britain; 14, New Hebrides; 15, New Caledonia; 16, Tonga–Kermadec

continental margins is the apparent seaward migration through time of the margin and the associated trench.'

(i) The mechanism by which a slab subsides is probably, once again, vertical laminar flow (Fig. 19). Because the resulting tectonics can be wholly vertical they can be quite independent of any supposed sea-floor spreading. Talwani *et al.* (1971) have lent support to this view: 'Downward movement along the continental margin, however, cannot be denied, both from the geological structure and the large negative gravity anomaly observed . . . the trenches along orogenic margins may indeed be comparable to the back deep of an orogen. If so, they are of local significance and closely related to the tectonic evolution of the orogenic margin itself, rather than being the result of large-scale collision of plates driven from far away.'

(j) There is no need to advocate an ocean-floor spreading and conveyor belt theory for the formation of subductogens. Indeed, and as shown by Elsasser (1968) 'the forces that generate and maintain the trenches originate locally rather than by transmission from very far away'.

Opinions such as the following appear to be exaggerated, and are possibly wrong in principle: 'Thousands of kilometres of oceanic lithosphere have been subducted at the trench system around the Pacific Basin since the Cretaceous. At least 7000 km has been underthrust beneath North America, a similar amount beneath Eurasia, and probably about 500 km beneath South America and/or western Antarctica.'

(k) Ultimately, the subsidence of the distal part of the slab (by that time broken physically to small pieces) carries it down to regions where its chemical and physical state may be altered beyond present recognition.

(l) Fossil subductogens are found in orogenic belts that have been welded on to an adjacent continent. The circumvallations of Gondwana and Laurasia were made in this way during early Phanerozoic time.

(m) The data indicating *rise* of the mid-ocean ridge-crest and *sinking* in subductogens at the oceanic margins have been linked in the minds of researchers, and on the principle that 'what goes up must come down' the two distinct (and opposite) phenomena have become connected in theory by convective cells within the mantle. A much more normal linkage between sites of elevation and sites of depression is by simple tilting of the landscape (or sea floor) between them, without lateral transfer of crustal and subcrustal material in convective cells. There are, indeed, learned papers on 'the older the crust the deeper it sinks'; but these seem generally to be tied up with supposed sea-floor spreading, whereas tilting may be simply and easily achieved by vertical laminar flow within the materials of the outer Earth. No lateral transfer either of crustal or of subcrustal matter between the cymatogen and the subductogen is then necessarily involved. If it were, it could be accomplished (as in salt tectonics) by transfer at depth of mobile asthenosphere from beneath the sinking area to beneath the rising zone (Fig. 13). There need be no 'return flow' as required with postulated 'convection cells'.

Areas of ocean floor have been consumed by subduction, to the extent that no portions of ocean floor that are significantly older than Jurassic have yet been identified.

Descending slabs, the de-gassed residues of former basaltic volcanism, sink heavily down into the hot upper mantle where they are assimilated and dispersed, thus losing identity. Whether the material of assimilated slabs then makes its way, under supposed convection, back to the mid-ocean ridges is dubious, especially under the high viscosities prevailing below the asthenosphere.

Also, to begin with, the original rising phase consisted of *light* volatile-charged lava intruding from depth. Risen, it emerged and gave up its volatiles to the atmosphere and the hydrosphere. These volatiles cannot be recovered, they are lost for ever from the subcrust. Melting of a subductive slab does not replace the original lava, despite the statement of the geophysicists' creed that it is 'made mantle' again. If it be reliquified, it does not thereafter journey back to the mid-ocean ridge, but rises vertically directly above the melting slab, where it comes up as *andesite*, not basalt. Any supposed convective circuit is broken.

Sea-floor spreading is real, and is borne out by the patterns of geomagnetic stripes; but it is a one-way movement to either side from the mid-ocean ridge, and even that is not in streams but a lateral shift of zones away from the ridge. There is no return phase, and the apparent lateral shift of zones, with time, is produced by vertical movement on the diverging radii of the globe, itself expanding in a one-way operation which, for the moment we describe as: 'A general swelling of the mantle.'

Finally we note that the mid-Atlantic Ridge is a well-marked, vertically-operating feature; but there are no trenches nor subduction slabs to either side of that ocean. And the same relationship (good ridge, no sinks) exists between Australia and Antarctica.

Perhaps the whole concept of thermal convection currents is not valid under the physical conditions prevailing within the Earth-body, attractive though the theory has been to many geologists.

A model for subduction

A useful model of subductogens was put forward by Worzel (1965) who associated them with a transitional zone of faulting and crustal warping (approximately 200 km wide) about certain continental margins. From this model he obtained remarkable comparisons with (a) the Aleutian, (b) the Tonga, and (c) the Puerto Rico Trench, without any horizontal sea-floor spreading or conveyor belt convection (p. 69) but using vertical tectonics. In this connection it is perhaps significant that the deepest trenches are often marginal to the steepest coastlands.

No specific explanation was given for the original block subsidence; nor was there mention of any subcrustal source of energy. We have already noted that seismic surveys usually record the presence of a mass of anomalously light (levitated) mantle in such situations, beneath the heavier (de-gassed) oceanic crust; and that gravity reversal between crust and asthenosphere is almost obligatory in a system where the mantle is de-gassing beneath a relatively impervious capping crust. Such reversal of density gradient at the edge of the ocean basin permits, where the Earth's structure is agreeable, the edge of the ocean crust to sag into the asthenosphere, under the operation of gravity.

At depths where the deeper earthquake shocks occur temperature and pressure may approach but little below those required for the dissociation of water. Dry steam within the rock bodies may form the porous type of magma which Einarsson (p. 48) has noted as the primary agent carrying volcanism to the surface. Owing to the inclination of the Benioff seismic zone beneath the edge of the continental-type crust, this condition may (like the earthquake shocks) travel along the top of the slab until at higher levels it follows the vertical laminar structure up to the surface where it forms the volcanic arc, or if the change of direction takes place earlier (because the vertical laminar structure is better developed in continental crust) the magma rises through the orogenic mountain structures of the continental margins to very high levels indeed, e.g. the volcanoes of the Andes, and Mt Elbruz in the Caucasus.

So vertical tectonics appear to dominate the structures of subductogens as they do those of cymatogens; the deep-sea trench as much as the mid-oceanic ridge.

Major subsidence independent of subduction structures

All the ocean basins afford evidence of subsidence (amounting to hundreds and even thousands of metres) in areas far from land. In these there is no evidence of breakthrough of the crust. For example, deep-sea cores show that continuous accumulation of shallow-water sediments over prolonged intervals of time is not infrequently followed by rapid subsidence to much greater depth. Beneath these basins the oceanic crust is usually thin, only 5 km.

Enquiries have been made into the location and tectonics of large-scale basin subsidences and the occurrence of petroleum therein, e.g. Fischer (1975) and Kinsman (1975).

To accommodate these displacements, physical changes such as corrosion of the underside of the crust at the Moho, or chemical changes such as metamorphism at the same level have been proposed. Petrologists (e.g. Ringwood, 1974) have assigned an importance here to amphibolite-grade metamorphism (with incorporation of water into the amphibole molecule); and Falvey (1974) has remarked: 'Such amphibolites commonly show considerable crustal manipulation.' These chemical changes apart, lateral transfer of asthenospheric materials, under the local loads of an impervious crust of de-gassified, inert basalt, affords the most likely mechanism for the deepening of ocean basins over wide areas.

Chapters 2, 3, and 4 have emphasized the role of vertical tectonics in the Earth's crust particularly along the zones of greatest uplift and greatest subsidence. But one must bear in mind that every part of the globe — on the continents or in the ocean basins — provides direct geological evidence that formerly it stood at different levels, up or down, and that it is subject *in situ* to vertical displacements. These movements originate locally by changing energies in the mantle; and since the asthenosphere and mantle everywhere underlie and support the crust, it is logical that: (a) vertical tectonics should operate universally within the crust — upon a sphere; and (b) lateral migrations or subcrust should occur and be responsible for changing elevations of earthscapes above and below the sea (see 'property of subductogens' (m) above).

Chapter 5

Lateral crustal movements: continental drift and the hypothesis of sea-floor spreading

The advent of the continental drift hypothesis

To geologists in the early years of the twentieth century, combinations of up or down crustal movements seemed sufficient to explain the distribution of continents and ocean basins about the globe — although in the literature of the subject lurked some nagging references to remarkable resemblances of stratigraphic data between the various continents of the southern hemisphere. Edward Suess in *Das Antlitz der Erde* had indeed envisaged a vast southern continent in Palaeo-Mesozoic time and named it Gondwanaland. Large portions of this land mass were thought to have subsequently foundered into the Atlantic and Indian Oceanic basins, leaving as relics the present southern continents (South America, Africa, India, Australia, and Antarctica). This hypothesis of wholesale foundering was soon found to be untenable. Gravimetric and seismic researches were interpreted isostatically for instance, as showing the physical impossibility of sinking continental crust of density 2.7 to suboceanic depths where the basin floor was of contrasted chemical composition, and of density 3.0 or more.

This impasse was resolved by the hypothesis of lateral drift of the continents, permitting the southern land masses to be refitted into a Gondwanaland that was little greater than the sum of their present continental areas. The accuracy of this viewpoint was then proved on geological grounds by du Toit (1927, 1937) who, by his detailed personal knowledge of the lateral variation within the Palaeo-Mesozoic formations of southern Africa and of South America, was able, to estimate how close they were together in those times (Fig. 21).

But the drift hypothesis introduced into Earth tectonics a new viewpoint — large horizontal displacements of continental crust about the surface of the planet; and for this break-up and travel no clear mechanism was cited

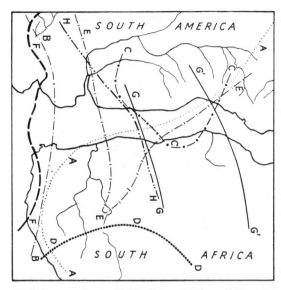

Figure 21 Stratigraphic, tectonic, and phasal comparisons between South America and South Africa. Siluro-Devonian Sandstones to west and southwest of AA; glacials unconformable to north of BB; Brazilian 'Coal Measures' (Bonito) within CC; South African 'Coal Measures' (Ecca) to east of DD; Ecca (blue-green) between DD and EE; Ecca (red) to northwest of EE; Gondwanide foldings FF; Gondwanide upwarpings GG, G'G'; Triassic unconformable to the north of HH (after A. L. du Toit)

beyond Holmes's (1931) exposition of thermally-controlled, subcrustal convection currents. This lack of known mechanism caused many geologists to reject the continental drift explanation; but the later accretion of data from postwar suboceanic research, coupled with belated acceptance of continental drift data called for synthesis in geologic thought; and this was attempted by Hess in 1960.

Sea-floor spreading

Building upon the concept of subcrustal thermal convection which had lain fallow since it was propounded by Holmes (1931), Hess visualized a subcrustal convection by which new basaltic magma rose from the subjacent mantle and injected the lithosphere along the crests of the mid-ocean ridges. This, he thought, made new crust, which was in turn thrust aside by further injection and created thereby a spreading of the sea floor to either side of the ridge crest — with separation and lateral drift of any superjacent continental masses. In certain instances (e.g. Atlantic) the present mid-ocean ridge was deemed to mark the former junction of the present continents of Africa and South America.

According to Hess's view, the conveyor-belt type of activity continued until the basaltic lithosphere became sufficiently cooled and densified to sink down into the mantle again. The site of sinking was supposed to be marked by a submarine trench in the ocean floor (see 'Subduction'). The basalt crust there dutifully sank, and at depth it was thought to be rejuvenated by accession of heat from the mantle. It then completed the convection cycle by transferring at depth from the trench site back to beneath the suboceanic ridge where it rose again to continue the cycle of sea-floor spreading. So, in theory, twin simple continuous convective systems existed to either side of the great suboceanic ridges, e.g. the Atlantic Ocean. This ocean, of course, has no lateral subduction trenches which, authority assures us, may be taken to indicate that the Atlantic has opened in width by the full amount of sea-floor spreading from the mid-ocean ridge.

In practice the pattern of the Indian Ocean basins (p. 153) is by no means as symmetrical; and the western half contains many islands of both continental and oceanic origin. The eastern Indian Ocean basin has few islands and the Indonesian island arc in the northeast sector is of Pacific type. The southwest branch of the Indian Ocean Ridge is not a normal cymatogen but a series of megashears. It has no crestal rift, nor is any geological evidence of sea-floor spreading associated with it (p. 152).

The East Pacific Ridge is quite askew in that ocean. It commonly lacks a crestal rift suited to injection from below, and at the northern end it is lost to view beneath the North American continent!

What is clear, of course, is that upon a non-expanding Earth only one ocean (the youngest) would be able to display a simple, bilateral symmetry. The others would necessarily suffer interference and encroachment upon their earlier features. Moreover, according to the stratigraphical history of the respective coastlands the Pacific is the oldest ocean, with the Indian and the Atlantic–North Pole oceans formed later in that order. Conversely, as all the southern continents drifted away from one another during the late Mesozoic, most of them could drift only into the Pacific (the original pan-thalassa) the periphery of which is almost ringed by the trenches of subducto-gens (Fig. 20).

Patently some careful explanations are called for, and perhaps the most puzzling problem of the relevant Earth mechanics is how segments of the lithospheric crust are caused to travel thousands of kilometres in diverse directions subhorizontally around the curve of the Earth.

Of great importance is the nature of the basaltic lower crust which, during the late Mesozoic when much of the continental break-up and drift occurred, must have been potentially eruptive upon a global scale, as is shown by the wide distribution of plateau basalts synchronous with the motion of the continents. This basaltic, lower crustal type is universal beneath the continents and ocean basins alike. Isotopic and trace element evidence suggests that it is derived as a product of fractional melting of the upper mantle on which it rests, sometimes with an intervening layer of gabbroic complexes and peridotitic rocks. But there is more. This upper mantle is abundantly charged with primitive volatiles at high temperature and low viscosity, which confer upon the melt an extraordinary state of mobility and vitality.

Hess's philosophy of sea-floor spreading was a high point in the interpretation of submarine bathymetry and tectonics; and the accompanying list of probable conclusions is formidable. It came at a time when research vessels became available to large numbers of geophysicists who took their instruments to sea and in the relatively short time of twenty years produced an incredible amount of new information on the form and nature of the sea bed, and the composition and structures of oceanic crust below it. Data poured out on gravity patterns and profiles, heat-flow rates through the Earth's crust, and geomagnetic reversal patterns within the suboceanic crust. The bottom sediments were plumbed and sampled by deep-sea boreholes. The newly-designed instruments for remote sensing of the physical properties of the Earth recorded it all in figures, formulae, and graphs, all of which required interpretation in terms of known geology. Geophysicists (most of whom had formerly rejected the idea of continental drift) rediscovered it, and taking over this piece of geological real-estate they erected on it a new field of study — plate tectonics (Chapter 7) (see Foreword). But before we are carried away by current enthusiasms let us examine some of the queries that have arisen on the sea-floor spreading theory.

Some problems of sea-floor spreading

By the new explorers of the ocean depths the hypothesis of sea-floor spreading was greeted with enthusiasm, and thereafter they explained most of their results in conveyor-belt terms. But despite much later information the concept remains assumptive and its mode of operation speculative: is such a convective system either true or necessary in nature? If there is such a system, is it operated solely by heat differences? And does it act continuously or spasmodically? A number of researchers have expressed doubt regarding the existence of such a pattern of thermal convection currents within the Earth, even allowing that the probable viscosity at the operating depth might not be too great.

Here are some of the queries:

(a) Convection is never a regular process. Stable convection cells of the size prescribed have nowhere been demonstrated. The zig-zag parts of a ridge–rift system are difficult to account for on an hypothesis of tensional spreading, and some other explanation seems required. Bullard states that convection does not normally happen along lines, and certainly does not happen along lines broken by frequent offsets such as affect the major oceanic ridges. He and others also note that on the world pattern not only the continents but also the ridges would need to be moving if plates of ocean floor were to form on both sides of a continent, e.g. Africa.

A much more satisfactory interpretation, for which we are indebted to Hast (1969) (see p. 35), is to regard the ridge as formed along the first of a pair of coupled shear directions, and the numerous transverse shears, abundant across all mid-oceanic ridges, as the conjugate system. This agrees with the analysis of seismic data (Hast); it restricts the crestal rift to vertical movement due to surface elongation on the arc–chord relationship of the rising arch; and it does away with the 'transform' faults prescribed by Wilson.

(b) Symmetrical injection and outflow of lava from the ridge axis is assumed; but it is not likely to be so in nature. The boundaries for the claimed magnetic polarity reversals are unlikely to coincide with natural outflows of lava.

(c) Calculations by Meyerhoff and Meyerhoff (1972a) of the amount of new crust injected at mid-ocean ridges did not tally with amounts of old crust consumed in subductogens. After all, the emission of magma and its solidification at the ridge crest involves more than cooling and change of physical state. Irreversible changes have taken place, and quantities of volatile activators have been lost to the atmosphere. What has to be disposed of in subduction zones is essentially dead end-product which cannot be reactivated merely by heating it up: it can only be melted. The *fons et origo* at the submarine ridge is essentially a vertical uprush of energy which is thereby lost. At the subductogen is a vertical sinking of 'inert' matter. In their opposite ways the two local processes are repetitive to finality. Why should there be horizontal transfer of dead residue between them?

(d) Later evidence shows, moreover, that in their present form the mid-oceanic ridges are, in part at least, major creations by relatively modern (Miocene and later) vertical uplifts, in which case they do not necessarily have connection with the continental drift (horizontal displacements) of late Mesozoic time. Moreover, in common with the volcanism associated with arches and rift valleys of the lands, their structure does not imply lateral extension of the Earth's crust (p. 38).

(e) Einarsson (1968) considered that the numerous faults transverse to the ridges (see charts published in *National Geographic Magazine* 1967 and elsewhere) are not a direct consequence of the postulated main convection currents; these faults, in sea-floor spreading language, were called transform faults and were due to abrupt changes in such currents — a state which Einarsson does not believe to be possible. These offsets have led on some diagrams to unrealistic representation of bathymetric forms. Instead, Einarsson regarded the whole structure of the Reykjanes region of Iceland, and of the neighbouring Reykjanes Ridge, as having been defined at its earliest stage by two conjugate shear directions operating in a global stress field (one direction is parallel to the Reykjanes Ridge and one parallel to the Faeroes Ridge (Fig. 22)), in that region of the Atlantic.

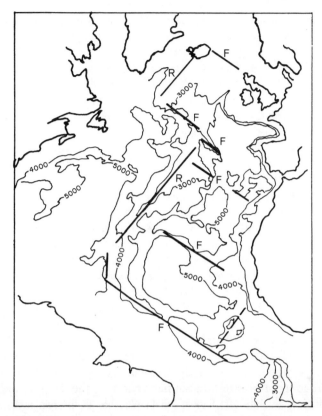

Figure 22   The main network of shear zones in the North
Atlantic as inferred from bottom relief and researches in
the Icelandic area. R = Reykjanes strike; F = Faeroes
strike (after T. Einarsson)

Observations by Hast (1969) in Iceland lead to some correlation of geomagnetic reversal patterns (Chapter 6) with stress fields in the crust rather than
with global magnetics (p. 76). To assess the stress fields it was necessary to
begin the study on some portion of a ridge which was above sea-level. As
the Reykjanes sea-floor Ridge is the classic area for the study of geomagnetic
polarity reversals in mid-ocean ridges, and Iceland is about the only landmass
arising from such a ridge, the Reykjanes Peninsula was peculiarly suitable
for examination (p. 78). Study of the fracture system there led to the conclusion that 'the Reykjanes Ridge represents a shear plane with dextral shear.
The alternate main elements of the zig-zag mid-Atlantic ridge are then taken
to correspond to conjugate shear planes. The observed mechanism of
earthquakes in fracture zones may be explained on this basis.'
    'The geology of Iceland also reveals a complicated history of that part of
the mid-Atlantic Ridge. At least the Reykjanes Ridge must have had the
same complex history, and a similar history is then to be assumed for ridges

in general. Assuming similar trends to those found in Iceland, one sees that linear magnetic anomalies can be accounted for in a general way without the assumption of spreading of the ocean floor.'

Over a dozen summers, Einarsson (1976) geologically surveyed the southwestern part of Iceland from Reykjanes Peninsula (10 km wide) as far inland as the Thingvellir and Langjökull areas (45 km). This is on the median active zone of Iceland marking the continuation of the active median zone of the Reykjanes submarine ridge. At the southern end of the Thingvellir Lake is an 8 km wide depression, wherein is an uninterrupted series of subaqueous eruptive rock strata not only within the depression but continued over the dolerite to either side.

'The denudation is on such a large scale as to suggest an age of many hundred thousand years. This is verified by the occurrence of reverse magnetization at a stratigraphically high level, setting here a lower age limit of 0.7 million years. *As these strata have been traced without a break right across the 8 km interval it can be demonstrated that crustal "spreading" has not occurred, or has been less than the often assumed 2 cm/year by at least two orders of magnitude, during the last 0.7 m.y.*'

Then, Einarsson (1976), regarding a postglacial flowage of lavas from the large shield volcano Heidin há, has written: 'A minimum estimate of its age is 5000 years. These lavas then tell us what tectonic movement has taken place on the prolonged Bláfjöll fracture line during a rather long time interval, and the answer is: none.' Also, of the median, active zone of Iceland where crustal 'spreading' at a rate of 2 cm per year over the last one million years has been widely acclaimed, Einarsson (1976) has delivered judgement: 'By the figure for speading in this zone, 2 cm/year as it is endlessly repeated in the geoscientific literature, the observable spreading ought to have been around 20 km. *This absence of spreading shows that the original, and largely still retained, explanation of the linear magnetic anomalies along the crestal zone of the Reykjanes ridge is wrong.*'

And again: 'Just before the last glaciation, the uplift phase of the Reykjanes peninsula was also active in this area, and a 17–18 km long SW–N arch was formed by uplift, with a maximum height (800 m) in the Hengill Mountain.'

The volcanic and tectonic activity in the median zone is considered to be the result of yield to an extensive crustal stress field, the phases for which express themselves somewhat differently, as can be explained in classical mechanical terms. These land-based data must lead to conclusions regarding interpretation of the Reykjanes Ridge — promised by Einarsson in Part II of his researches (1977b). Therein he examines also the seismicity pattern on the 1500 km wide submarine ridge and interprets it as due to strike-slip and dip-slip fracturing 'in the uppermost 3–5 km thick crust' (cf. Hast, 1969). The linearity of magnetic patterns he then attributes to the magnetization vector of the affected rock becoming aligned with the shear movement leading up to the shallow earthquakes of the ridge areas, and so creates the pattern of the magnetic anomalies. In support he quotes the Stardalur anomaly (on land) which is due to *altered* basalt, but has ten times stronger magnetization than is usual in basalts.

In a general statement on the effects of stratigraphic burial and non-hydrostatic stress he 'considered, in particular, magnetization in the light of the fact that magnetic or domain bonds are much weaker than lattice bonds, and inferred that the former have yielded in rocks as young as 60 m.y. and so changed the magnetization.' Such viscous magnetic change with regular new orientations, induced under non-hydrostatic stress he regarded as possible for oceanic basalts. They are regarded as the result of lattice changes; and extending the argument, 'Such boundaries are, in significant cases, mainly a function of depth, rather than of rock type. This is shown by horizontal boundaries right through the roots of folded mountains like the Alps. A careful study of the Sierra Nevada shows remarkably regular systematic boundaries going horizontally through the vertically layered granitic axial zone of the mountain chain and ignoring the otherwise complex structure of the chain. All the boundaries, to the Moho at 50 km depth, must have been *formed* by some process *after the orogeny*, possibly a long time after it.' He suggests: 'It must be a jump to high-pressure polymorphs . . .' and we may 'anticipate here that the Moho means a change to high pressure *nanocrystals*, which automatically create a far higher local pressure around themselves than corresponds to the lithostatic pressure at their depth.'

(f) Argument from the study of geomagnetic reversal patterns in rocks have generally led to the conclusion of smoothly continuing, slow spreading of the ocean floors. The geological record, however, is one of 'fits and starts' with short tectonic episodes followed by prolonged intermissions of relative quiescence. The geomagnetic record is more in accordance with movement of the patterns through the lithosphere than it is of magnetism 'frozen' into individual rock systems which then have to be moved (p. 76).

Further discussion of the nature of geomagnetic polarity reversals is deferred to Chapter 6.

(g) Half the seismic activity of uparched sea floors is narrowly associated with the rift zone and is shallow in terms of crustal origin. Furthermore, it is associated with fracturing (Figs 22, 42). The other half is more widely scattered and shows no special association with superficial features. It perhaps reflects the broader, cymatogenic growth. As this is true of both land and sea-floor areas, and as crustal spreading is not evident about the land rifts, perhaps it is not real upon the sea-floor either.

(h) Elsasser suggests that a form of convection occurs, but in reverse (a rather similar conclusion was reached by Hast); and Elsasser assesses the situation: 'there is no actual evidence for convective motion in the upper mantle caused by primary heating from below.'

(i) The cymatogenic nature of the principal suboceanic ridges under stress is supported by intermittent elevation of the ridges themselves during Cenozoic time as evidenced by Cenozoic sediments on various parts of the ridges.

Cifelli *et al.* (1966) for instance found upper Miocene foraminifera in the crestal rift of the mid-Atlantic Ridge, and Russian authors also reported 'Miocene sediments . . . near the Atlantic fault in the immediate vicinity of the ridge axis.' Such observations show that the rift arching and faulting was (like the Albertine Rift of Africa) active at that time. Lavas in the same situation have also been dated as 8.5 Ma and 18.6 Ma respectively. Rocks recovered from Mt Cobb on the crest of the South Pacific Rise (Heirtzler *et al.* 1968) have been dated as 29 Ma.

Facts like these tell strongly against the idea of sea-floor convectional spreading because only subrecent rocks should appear about the ridge crest (p. 81). They agree very well with data for the Miocene elevation of cymatogens with rift valleys upon the lands.

(j) Although several kinds of alkalic basalt and ultrabasic rocks have been brought up in dredge hauls from submarine ridges, most of the 'new crust' is tholeiitic basalt (rich in alumina, low in potassium). The lavas of the African rifts (rich in potassium and sodium) are about as far removed from these tholeiites as is petrologically possible.

(k) Several authors have drawn attention to the apparent lack of disturbance in the sediments of the ocean basins, and remarked that such regular horizontality over such vast distances is surely not in conformity with the concept of convectional spreading (see Chapter 9). The extent of Cretaceous sediments is so vast as to indicate that most of the present oceanic area was already in existence at that time. In other words, the late Mesozoic fragmentation of Gondwanaland was followed by very rapid dispersal of the southern continents, and relatively little drift in Cenozoic time (see Chapter 10).

(l) On the first voyage of the *Glomar Challenger* a special test programme of drilling at sea was carried out to test the problem of sea-floor spreading in the Atlantic. A discovery of old sediments in the middle region of the ocean basin would spell disaster for proponents of sea-floor spreading. The results were summarized by Cox: 'This cruise confirmed beyond any reasonable doubt that the Atlantic had formed by sea-floor spreading — pre-Mesozoic sediments were absent; the sediments were systematically older away from the ridge; and the age of the sediments lying immediately on the basalt basement were consistent with the time scale of magnetic reversals in the South Atlantic. A more direct and compelling confirmation of sea-floor spreading at a nearly constant rate cannot be imagined.'

Reference to the original report, however, shows that sites 21 (28° 35′ S, 30° 36′ W) and 22 (30° 01′ S, 35° 15′ W) on the Rio Grande Plateau yielded late Cretaceous rocks far out in the Atlantic. Rapid Mesozoic drift is indicated. Equally remarkable was the absence of any deposit between mid-Eocene and Pliocene at site 21, and the same gap but with upper Oligocene to lower Miocene present at site 22. Taken together, these records indicate not merely an intermission of non-deposition, but active erosion at one time, i.e. local destruction of the sedimentary record. There is also a shorter

(Middle Miocene) absence of sediment in cores 14–19 on the mid-Atlantic Ridge. The region evidently underwent elevation and submarine erosion at least once during the Cenozoic.

The original assessment by Cox, based upon a single area of the Atlantic, was very soon contested by many new records. Jurassic and Cretaceous rocks are quite widespread near the crests of ridges, and palaeozoic fossiliferous rocks (even Cambrian) now cause no surprise if encountered in the crestal situation.

Several authors have suggested non-uniform Atlantic spreading on the basis of heat flow, discontinuities in ridge topography, and 'amplitudes of magnetic anomalies' since the original suggestion of irregularity in Atlantic spreading rate was made by Ewing and Ewing in 1967.

On the flanks of the Atlantic Ridge, to about 30° north and south, mid-Miocene red clay is common in the deep-sea drillings. This is the time when the Mediterranean dried up, and all North Africa was desiccated (in the Trade Wind belt). Uplift of the ridge was strong in the Pliocene and Pleistocene.

Most of the phenomena noted by Cox are equally explicable by normal overlap of sediments upon the flanks of a ridge which occupies one-third of the width of the ocean, coupled with minor up and down movements of the ridge itself. Moreover, the sediments are mostly derived from the lands, not from the centre of the ocean.

(m) Incidentally, basins to either side of the mid-Atlantic Ridge have deepened intermittently since the continents separated. Such periodic deepening of lateral basins is indeed general, and appears to be synchronous with uplift of the boundary continents which are progressively tilted seawards (Chapter 10). The interfingering of denudational and sedimentary records in this coastal zone provides direct correlation of continental and oceanic histories of exceptional value (Chapter 11). They show, amongst other things, that global tectonics have been a 'stop and go' affair, with worldwide correlations, and that the often quoted 'average rates' of sea-floor spreading are no more reliable than other 'averages' that have sometimes been adduced in geology.

Some writers have sought accommodation. Thus recently, in a paper entitled 'Steady plate motion and episodic orogeny and magmatism', Pitcher (1975) after long study of the large Peruvian magmatic cycle (granitic) in which he noted the episodic nature of geologic events in that region, wrote: 'Nevertheless continuous activity best fits the model of sea-floor spreading, subduction and deep melting and, appropriately, the rates involved are of the same order, i.e., several cm per year for the opening and emplacement of magma at oceanic ridges and for the uprise of magma pulses to form batholiths at continental plate margins.'

But there can be no accommodation between basically opposed data, and discontinuities in sea-floor spreading were discussed by Vogt et al. (1969) who listed the following types: (1) total stoppage; (2) shift in location of the rift axis; (3) change in the rate of spreading; (4) change in the direction of spreading; and (5) change in the mode of spreading.

(n) The Antarctic continent is almost ringed about by submarine ridges. Sea-floor spreaders have thought that these spread outwardly in all directions. But not all are of the same age.

By the end of the Mesozoic most of the original impetus of continental fragmentation was spent, and several land masses had arrived near their present geographical positions. The areas still active during the early Cenozoic were: (a) splitting of the India–Australia–Antarctica fragment of Gondwana with the northward drift of India into Asia, the drift of Australia–New Guinea into Indonesia, and the southward drift of Antarctica into its present polar position — all these movements were completed by the Miocene; and (b) northward movement of Greenland with formation of the Norwegian and Barents Seas with much accompaniment of plateau basalt effusions.

Other doubts on sea-floor spreading were raised by Bullard (1968), Carr (1968), Maxwell (1968), van Andel (1968), Watkins (1969), Nafe and Drake (1969), Beloussov (1969), Mantura (1972) and Bonatti et al. (1975) who found Jurassic and Cretaceous at the ridge crest. There is also a long list of Palaeozoic rocks dredged from certain ridge crests for which the bibliography should be consulted.

Perspective was added by Dietz (1973): 'With convection one can do almost anything as the entire process is wonderfully amenable to mathematic manipulation. And there are many modes of convections — toroids, plumes with thunderheads, helixes etc., all of which can be readily explained by arm-waving which conjures up explicit models.'

To this author 'sea-floor spreading' expresses the action admirably; but he sees no evidence for 'conveyor-belt' technology. Instead, he comprehends a general enlargement of the mantle body within the Earth.

Horizontal stress fields in the Earth's crust have undoubtedly changed in magnitude and direction of shear during geologic time, as may be inferred from modern measurements. An extensive survey by Hast (1969) has shown a close relationship in measurements of the existing state of stress in the upper parts of the Earth's crust over large areas. Thus: 'Measurements performed in the Atlantic area in coastal zones, on Iceland and other islands seem to show that the Atlantic ocean floor behaves *as a rigid plate** stressed by horizontal compressive and shear forces, in all probability emanating from the zone of contact between ocean floor and continental crust. The mid-Atlantic Ridge and the numerous east–west fracture zones in the ocean floor seem to represent an orthogonal fracture system probably deriving from horizontal shear stresses in the floor.'

The role assigned to the continental–ocean floor contact at the sides of the

---

* Italics mine — *Au.*

ocean basins should be noted; and also the definition of the ocean floor as 'a rigid plate', not as a semi-fluid capable of flowage under stress.

Hast's first measurements were made in Fennoscandia, later the work was extended to Iceland, Ireland, Portugal, Nubia, Liberia, Zambia, and British Columbia until about 20,000 readings (50 to each borehole) had been made. Among all these recordings *not one* pointed to a state of tensile stress: 'All have indicated horizontal compressive stresses in the crust, several times the magnitude of the stresses due to the weight of the overburden at the same level; zero stresses in the horizontal direction have been found only in bedrock containing open fractures and at a few places where the shear stresses are extremely high.'

These observations, which might have been expected for a spherical Earth subject everywhere to centripetal gravity (Fig. 15), throw doubt upon deep-reaching tensile operations beneath the elevated crests of cymatogens, and sundry operations formerly thought to be carried out by sea-floor spreading.

At all areas observed, the magnitude and direction of the stress field was similar, and this applied frequently to conditions today and in the geologic past. 'The maximum pressure acts at 45° to the direction of the mid-Atlantic Ridge which means that the direction of the maximum horizontal shearing stress in the bedrock in this area coincides with the line of the ridge; this suggests that the (mid-Atlantic) ridge may have its origin in shearing stresses in the crust.'

This research over a wide area gives the positioning of the mid-Atlantic Ridge hereabout as decided 'by horizontal shearing in vertical planes', a structure compatible with injection of mantle material of suitable levity to rise up and make intrusions along the shear planes.

Hast's stress analysis applies to the equatorial dislocations of the mid-Atlantic Ridge and shows that the stress conditions to form the offsets between long east–west fractures parallel those of the North Atlantic. He noted also that the direction of maximum stress in the coastal zone of Liberia coincides with that of the fracturing of the Atlantic ocean floor between South America and Africa.

Hast then compared the fracture zones off the western coasts of North America, and considered that they too exhibited a joint system that might be of the same type as that in the Atlantic, and due to 'horizontal shearing forces acting in the ocean floor in the zone of contact with the continent'.

These researches demonstrate the presence of a rigid sea-floor plate carrying the same stress patterns over vast distances. When these data are combined with the rarity of shallow earthquake foci on the sea floor away from the mid-ocean ridges and the continental margins, and the demonstration that the mid-ocean ridge is a seat, not of tensile stress, but of horizontal and vertical shear components upon vertical planes parallel with the axis of the ridge (Figs 14, 15), the hypothesis of mid-ocean ridges as extensional areas horizontally, stands bereft.

The subductogens

Exponents of sea-floor spreading have postulated that the deep trenches found at some of the ocean boundaries are places of subduction where ocean-floor crust which has migrated like a conveyor-belt from the mid-ocean rift zone has cooled and sunk back into the subjacent mantle. But the distribution of trenches about the Earth (Fig. 20) does not support this view. Nafe and Drake (1969) have pointed out, for instance, that the Atlantic basins have no general system of marginal trenches or sinks, and if the Atlantic crust, spreading both ways, takes the continents with it, what problems are thereby created in the Indian and Pacific basins (e.g. in geomagnetic polarity anomalies)? The Puerto Rico trench is separated, in part, from the Atlantic floor by the Barbados Ridge, and the trench must be very young for the tectonic history of the island of Puerto Rico crowning the inner scarp of the trench is reflected in the sea-floor sediments resting on the outer side of the trench.

The other Atlantic trench, South Sandwich, is even more puzzling. Although its lonely position in the middle of the South Atlantic is probably connected with that outpost of South American geology, South Georgia, both the island arc and trench of South Sandwich are very young, indeed currently active! The Indian Ocean has no marginal trenches either,* and while the Pacific is girdled east, north, and west by clear trench forms this is not a disjunctive ocean but the realm of pan-thalassa into which the continents have crowded from all sides (unless indeed the Earth has expanded considerably during Cenozoic time, as suggested by Carey (1976)).

The presently popular conveyor-belt mechanism for continental drift falls flat on the simplest geographical examination. There is no proved consumption of ocean-floor sediment in any trench, and none of the modern trenches seems to be older than 10 million years. Some of them are much younger than that, as the Aleutian trench, which is between the ocean floor sediments and the crumpled sediments of the *inner* wall.

Coasts which were not original Gondwana or Laurasian coasts have no trenches. The Atlantic and Indian Ocean margins have no subduction zones, and sea-floor spreading by 'conveyor-belts' cannot be seen to operate without Cenozoic expansion of crust in these regions.

On geomorphological and stratigraphic evidence (see later chapters) it seems likely either that *few* trenches existed formerly, or that trenches existed in quite different places from those at present and that the sites of these have been obliterated. So what of the *average* rates of sea-floor spreading through the Cenozoic that are so often quoted? Recent DSDP data from the North Pacific suggest a limited amount of activity since the Eocene, 'with a resumption of underthrusting in the late Cenozoic'; and this agrees with the geomorphic history of the adjacent lands (p. 158).

In the South Pacific Menard has noted, too, that there is no evidence of

---

* The Indonesian trench belongs to the Pacific system.

the large quantities of sediment which continuous sea-floor spreading would have swept up to the margins of trenches for disposal by subduction.

So the nature and mode of occurrence of deep trenches along Pacific Ocean margins only tells *against* the conveyor-belt hypothesis rather than for it (p. 82). Though subduction is prescribed by geophysicists as the garbage disposal system of sea-floor spreading (p. 82) no rates of resorption seem to have been calculated; and disposal is itself necessitated only by prior assumption of conveyor-belt sea-floor spreading!

This unsatisfactory situation calls for a new start on the problem of horizontal crustal displacement. The outstanding fact from which the enquiry can begin is that the best actual demonstration of such horizontal displacement is the Mesozoic and later, dismemberment of Gondwanaland, an event probably unique in geological history. We then consider the vicissitudes which overtook the daughter continents in their antipathetic Mesozoic flight, and finally consider the quite different question of the extent, if any, of Cenozoic continental drift. This investigation must be undertaken on Gondwanaland because Laurasia in the northern hemisphere displays no comparable disruption.

The break-up of Gondwanaland

The mid-Mesozoic fragmentation of Gondwanaland (Fig. 2), followed by dispersal of the modern continents to great distances, is demonstrated to geologists by:

(a) comparison of the Palaeozoic stratigraphical sequences (of continental facies) in the several southern continents;

(b) by source areas of these sediments being so often beyond the present continental margins; and

(c) by palaeoclimatic interpretation of the Palaeozoic continental sequences which shows how Gondwanaland moved through a series of climatic zones consistent with latitude that are similar to the climatic girdles of the present day.

So large was Gondwanaland that it often spanned several climatic zones at one time. Thus as it drifted (westward*) across the south pole warming up was evident in South America while the Dwyka glaciation was at its maximum in South Africa, and Australia, coming from the east, had not yet entered the polar circle, so the glaciation dragged on there into the mid-Permian (King, 1962/7).

By the beginning of the Mesozoic much of Gondwanaland was established in the southern tropics, and the formations of the Triassic are wind-blown desert sands (Botucatu in Brazil, Cave Sandstone in Africa).

From the coincidence of late Palaeozoic girdles with present global climatic

---

* In the vicinity of the pole, of course, all poleward movements are south and all movements away from pole are north. West and east here refer to parts of Gondwana left or right in the reconstruction Fig. 2.

zones it is evident that the Earth then spun upon the same axis inclined at the same angle to the ecliptic as today. Continental drift, not polar wandering, clearly was responsible for the climatic changes recorded in late Palaeozoic–Mesozoic stratigraphy. To a Gondwanalander this is an oft-told tale, told by the rocks of Gondwanaland about him, part of his life and environment whether he be in Paraná, Karroo, Gondwana, parts of Australia, or even visiting along the great trans-Antarctic mountains of the Ross Sea region.

Laurasia and Gondwana probably existed as such from at least Proterozoic time. Prior to then the two supercontinents appear to have been accreting (Fig. 1) possibly as twin polar land masses. At the time of their collision (p. 6) both had a converging westward drift, possibly global.

When Gondwanaland disintegrated the fragments (now the southern continents) flew apart in all directions implying a centrifugal force beneath Gondwanaland, not a global convection system. As we noted (p. 6), the disintegration was conceivably triggered off by a glancing collision with Laurasia (Fig. 3),* but as the other pieces took up individual motion the evident repulsion of so many bodies, each from the others, is surely an indication that the parent supercontinent was already under a strongly disruptive stress field. The stress, and the forces giving rise to it, remained in control for the rest of the Mesozoic as the Gondwana fragments continued to drift farther and farther apart.

But the original impelling forces had faded by the Cenozoic, and as we shall see (Chapter 10) flight slowed down, and over all the continents quiet conditions prevailed through the early Cenozoic and widespread planation developed in all the several southern continents (p. 6), even those of the India–Australia–Antarctic assemblage which continued their drift individually until the Miocene.

The Miocene period saw a revival of tectonic activity. Uplift of all the continents occurred with geologic synchroneity. Over all the larger land masses of the globe a characteristic rolling landscape developed. At least two continents (Australia and Antarctica) experienced renewed drift as they split and travelled in new directions, until Australia drove into southeast Asia where New Guinea crumpled up the Banda arc (Fig. 3(d)), and Antarctica travelling in the opposite direction took up station at the South Pole and began the refrigeration which continues until the present day.

Further tectonic activity followed by denudation began at the opening of

---

* The first part to break away from Gondwanaland was Arabia (in the late Precambrian), followed in the Permian by the block of Iran–Afghanistan, which moved northward during the Permian, a time when the emission of widespread basalts in Siberia showed the presence of active levitated mantle beneath Laurasia. This was pointed out in King (1973). In 1974 von H. Forster considered that 'Iran is assembled by fragments of Gondwana, the margin of Eurasia and remnants of ocean floor.' Also, with Baker and Soffel he pointed out that the late Precambrian and Lower Cambrian rocks of Iran and India were related to the same geomagnetic pole, and hence at that time Iran was part of Gondwanaland. Madagascar became separated from Africa about 100 Ma ago during the late Jurassic — early Cretaceous periods but it did not go far and has moved in concert with Africa ever since.

the Pliocene, the coastal plains and interior plains of which form the most widespread landscape in all parts of the globe. By deformation, they indicate the prevalence of global tectonic controls, dominantly of uplift of the lands and subsidence in the ocean basins. But though the fundamental control is global (King, 1962/7), and is vertically similar in all continents, yet the continents behaved with freedom as separate entities already long-established in their present sites.

All these changing circumstances and ensuing chapters of landscape development (p. 184), each of them worldwide (King, 1976), requires explanation under any hypothesis of Earth development during post mid-Jurassic time. The sea-floor spreading dogma, offering slow continuous action, just does not meet the various and varying requirements of what proves to be a very ingenious earth.

How horizontal crustal movement may occur on a large scale

Another model of mantle–crustal activity is now suggested which:
(a) does not require the conveyor-belt concept;
(b) is not operated solely by thermal differentials but takes into consideration mass differentials of sundry mantle constituents also;
(c) can be operative locally, regionally, or even at times globally, depending on the relative local strengths of the crust, and the degree of mantle activity below.

The basic assumptions are no more than those usually accepted among geologists. First, it is assumed that the Earth's crust beneath the ocean-floor sediments and the sialic continents alike is basalt, usually tholeiitic. By its nature the basaltic crust may be relatively impervious, but where it is locally piped or fractured, conduits are provided that could be recognized, whether on land or sea, as 'hot spots'.

Second, it is assumed that the basaltic crust rests upon an asthenosphere or upper mantle zone stored with primitive and radioactive heat, and richly charged with volatile constituents, cf. du Toit's (1937) 'paramorphic zone'.

Third, it is assumed that these volatiles tend to escape upwards by levity. De-gassing of the mantle has progressed, indeed, throughout geologic history, and the sum total of its activity has provided the atmosphere and hydrosphere which surround the solid Earth.

The Earth being a sphere this outward migration of heat-bringing volatiles into and through the Earth's crust must be a global process (p. 16), measurable approximately by heat flow anywhere upon the Earth's surface. It is a simple leakage from the interior by levity, not requiring any convectional system. Volcanism from the upper mantle is thus latent everywhere. The main property of these active materials is that they are omnipresent, available always and everywhere directly below every point upon the Earth's surface. Egress generally depends upon the availability of passageways through the crust such as faulted or fractured structures leading upwards, e.g. plate boundaries (p. 117), megashears, or certain types of continental margins.

Where the passageways are elongated at the surface, plateau basalts or suboceanic ridges could be typical. The zone of increased heat flow evident along the crests of certain such ridges is very narrow indeed, indicative of vertical escape only, and at relatively short distances from the active crest the usual slow rates of heat escape through the sea floor are resumed.

The presence and abundance of gases (mostly volcanic steams) in these situations seems to govern the extent and activity of tectonic manifestations more than the mere rise of temperature. Similarly, volatiles and viscosity seem to govern mantle activity and the rise of volcanism more than do melting points.

A solid rock, for instance, is incapable of motion under its own volition. Heat it, and the rock becomes plastic and melts, when it becomes capable of deforming under applied external stress. When there is addition of volatiles such as magmatic gases the rock takes on vigorous activity of its own. No longer does it react passively to external forces, it becomes a positive power on its own account, deforming and opening up its surroundings instead of adapting to them. Petrologists have even distinguished 'fronts' (both physical and chemical) in large bodies of igneous rock, indicative of 'bad weather' in the substratum.

Volatiles (especially at high vapour pressure) help to lower the melting point of the host. They can actively carry heat from place to place in a melt. They markedly lower the viscosity of the melt. And finally they greatly reduce its density. All these properties assist or require that the volatile-charged magma under the static load of the overlying crust shall rise from the depths up any channel available and make its way upward through the crust. Alterations in composition (usually with increase of silica) during the ascent, and progressive decrease of density as the magma rises and comes under reduced pressure are all primarily functions of the volatile content. As an example of their action: in basalts a high partial pressure of oxygen gives rise to minerals in the ferric state, a low partial pressure to ferrous state minerals.

The easiest way out will always be upward, under levity, in the manner of salt tectonics. Good conduits permit the extrusion of voluminous plateau basalts: where passages through the crust are more constricted 'tear drop' reservoirs of magma make their individual ascents. At no stage are conditions likely to favour lateral dispersal with conveyor-belt travel to great distances, though small convective cells are known to have occurred in some large basic intrusions. Heat is evidently less important in igneous activity than the heat bringers which themselves become more acid on the way up (a) by assimilation, and (b) because much of the juvenile steam carries silica in solution, even to excess so that the issuing basalts are amygdaloidal with agates, etc. The final congealing of the lava signals the loss to the atmosphere of the attendant steams with their load of other chemicals.

Once action has begun by ascent of levitated mantle into localized regions of the crust, enormous reserves become available (Fig. 13) by subcrustal migration *towards* the volcanic region under the compulsion of static load exercised by the crust itself, and neighbouring regions tend to subside gently. Thus subsided ocean basins come to lie on either side of uparched submarine ridges wherein are large basaltic intrusions and effusions of narrow total width. Action may continue until the local sources of subjacent mobile material are exhausted and indeed most operations are of no more than regional scale.

Where the mantle energies are confined not beneath a relatively thin oceanic crust but beneath a hyperfusible blanket of much thicker and stronger continental crust, escape of the volatile-charged upper mantle may be inhibited. Pressures and to a lesser extent temperatures* may be expected to rise gradually as the volume of levitated mantle increases.

For a prolonged interval of time the continent may be expected to sag gently, especially towards its centre which may thus become dished and serve as a focus for continental-type sedimentation. This was the stage reached by the Palaeozoic Gondwanaland at the time when the Karroo system of South Africa was laid down near the centre of Gondwanaland, with equivalent basins in South America and India. The earliest member of the succession was the Carboniferous 'Dwyka' embodying polar glacial beds. The youngest is the Jurassic Gokwe beds of Rhodesia, with *Massospondylus*, and again with stratigraphic equivalents in South America and in India.

As the supercontinent subsided, tensional fissures leading upward would form at the base of the sagging crust. Advantage would be taken of these for the potent magmas of the upper mantle, with their volatiles, to begin a large scale invasion of the overlying supercontinental crust. Widespread dyking (with sill intrusion into suitable rock formations (shales)) then reversed the continental sagging and instead began domings of both the base and surface of the supercontinent. Basic and ultrabasic plutons were then intruded, e.g. in Griqualand East. This is the story of the Karroo dolerites, where du Toit estimated that the volume of dolerite injected beneath South Africa alone was between *50 and 100 thousand cubic miles*; but similar dyking, with additional sills and plutons is known in Tasmania and in Antarctica. The crust of Gondwanaland became engorged with levitated mantle, and because of the new doming was placed under a set of centrifugal forces (partly gravitational) with each sector of its periphery tilted outward (or forward) ready to fly apart. In a manner of speaking, it was 'overripe', and at this stage it collided with Laurasia!

This is the only example of centrifugal fragmentation of a supercontinent known in geologic history. Maybe there have been only two such supercontinents, Laurasia and Gondwana, in geologic history. But it is the story told by the geological formations of all the parts, even in their present dispersal.

---

* Lesser, because some heat may be lost by conduction.

To continue, as the magmas finally reached the surface in the mid-Jurassic they poured forth the immense floods of plateau basalts covering hundreds of square kilometres in Brazil, in South Africa, in India and in Antarctica. This was the moment of disruption (Fig. 31).

Henceforth the present southern continents were on their own. Each daughter continent inherited a leading edge of fold mountains that had formed part of the Gondwana circumvallation, each was tilted forward in the direction of travel, and each had to supply its own motive power. Each began, of course, with a reserve of levitated mantle below it, thrusting upward against the outwardly tilted base of its crust. The oldest continental basement rocks are often exposed by Cenozoic erosion over those parts of each daughter continent which originally lay towards the centre of Gondwanaland.

Each continent rode as it were upon a cushion of levitating mantle. The power source might be expected to fail ultimately, but to begin with each continent was powered like a rocket. Later power surges are indicated by further outpourings of plateau basalt in early Cretaceous time (Brazil and southwest Africa) and late Cretaceous to Eocene (India), so that propulsion died down by the end of the Cretaceous. Australia–New Guinea experienced a belated revival of propulsion and drift in the Miocene, when it had much volcanism in the east and caused the intussusception* of the Banda arc at Sulawezi (p. 7). Antarctica, which was involved in opposite movements, took up its present polar position simultaneously in the Miocene. But from the remaining southern land masses there is, as yet, no compelling *geological* evidence of Cenozoic drift, only of vertical displacements *in situ* (p. 44). Certainly there is no orthodox geological evidence of Cenozoic subcrustal convection currents, or of sea-floor spreading.

Instead, the record in all the southern continents is of quiescence and stability as though de-gassing of the mantle had died down. Vast planations by denudation are typical of the early Cenozoic continents (Chapter 10).

Late Cenozoic time displayed renewed activity. On both the lands and the sea floor are huge megashears. Cumulative movement along the San Andreas fault since the Miocene is 1400 km, and on the Alpine fault of New Zealand is more than 500 km. On the sea floor are even larger displacements along the Clipperton and Eltanin fracture zones of the eastern Pacific. On some of these the horizontal shear movement passes through the full thickness of the crust, and though vertical displacement usually appears small in comparison with the horizontal, in places the faults involve elevation of ultrabasic masses from the mantle below.

The phenomena accompanying the disruption of Gondwanaland and the centrifugal dispersion of its fragments, with distinct mid-Jurassic, late Cretaceous and Miocene episodes of drift, and quiet intermissions of stability between, form the prescription which must be fulfilled by tectonicists. It is

* Intussusception = turned inside-out like the finger of a glove.

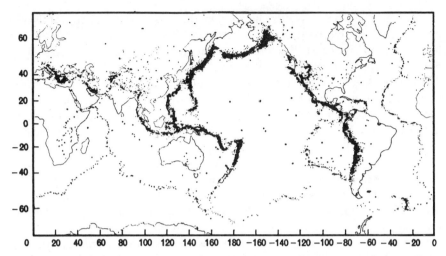

Figure 23   Shallow earthquake foci (<99 km) for the period 1961–7. The distribution is: (a) circum-Pacific; (b) Alpine–East Indies; (c) suboceanic ridges and rift valleys in descending order of abundance (after Fairhead and Girdler, 1971)

a problem immensely grander than the opening of the North Atlantic, which has received a disproportionate amount of attention.

The small volume and relative youth of flat-lying sediments in the Peru–Chile, middle America, and Aleutian trenches also negates the conveyor-belt sea-floor spreading model for Cenozoic time. Sea-floor (plate) movements, on the contrary, must have been widely operative, mainly during the late Mesozoic. So let us examine once more the concept of widespread horizontal (plate) displacements in the oceanic and continental areas, probing to find a cause and a mechanism for movements of this type, and remembering first that Hast (1969) concluded from seismic measurements that the Atlantic sea floor may act as 'a rigid plate' we enquire what other areas of the ocean floor appear to have evolved in such a manner? The most important example is the northwest Pacific, described by Heezen and MacGregor (1973) who deduced from its sedimentological record that a huge area of sea floor has there travelled platewise from the southern hemisphere, across the equatorial zone of calcium-rich faunas into the present silica-rich province of the northern hemisphere deposition (Chapter 8).

Continentwise, there is also the creep of Africa towards Europe with the transfer of bits of Gondwana stratigraphy on to Italy and maybe Greece, that was part of the collapse of the Tethyan sedimentary basin, an event that led to the Alpine orogeny of southern Europe. An even clearer demonstration is that first Arabia, then Iran–Afghanistan, Baluchistan, and finally India, all of them former portions of Gondwanaland, have crossed the equator and attached themselves individually to the southern flank of Laurasia.

Laurasia itself has travelled north. On the west coast Vancouver Island and Baja California have separated from mainland America and advanced northward up the coast. Tectonic movements on the Laramide front (Fig. 37) are northerly at the heart of North America.

Then the Arctic Archipelago north of Canada is part of a continental girdle that, followed eastward through northern Europe and Siberia, encircles the boreal area between 60° N and 75° N. This girdle (Fig. 36) has no equivalent in the southern hemisphere: its place is taken by the Great Southern Ocean. Instead, this girdle has the aspect of a region that, beginning with expansion as in Fig. 31, has been caused to reverse its early expansion and has latterly shrunk in area. In doing so it has gathered its component lands together like pack-ice. The effect of moving land blocks of fixed size into an area of converging meridians does not reduce the size of those blocks but eliminates the seaways between them (McElhinny and Brook, 1975).

Who can doubt this mighty drift of lands and sea floors *northward* in high northern latitudes, following an earlier phase of normal global expansion? The change was brought about very rapidly, almost instantaneously, as one or other of the geomagnetic poles swept closely past the equatorial position where the mantle streams were adapting to a new, superficial location and were chemically in a form that was adaptable to change. At present we do not know the cause of change; but fluctuating geomagnetic influences offer change of satisfactory speed and control over a sufficiently large area. The total area of the global surface under the control of southern hemisphere pattern at the present time represents 90% of the global surface. The remainder is merely a small area round the north pole. Apparently this inequality has existed for a long time, probably the whole of Phanerozoic time at least.

The nature of conflict between the equatorial mantle streams of the two hemispheres

The mantle streams of the Earth are, we have remarked, radial in both hemispheres, and they move towards regions of greater space as they reach the surface, so that they do not meet in direct head-on conflict at any stage, though we observed surface features which suggest that while northern hemisphere streams develop a southerly component of slow flowage, *the southern hemisphere streams develop a strong northerly component* of flow which is clearly seen in the drift of fragments of Gondwanaland — Arabia, Iran and Afghanistan, Baluchistan and India, all of which have crossed the equator and attached themselves to Laurasia. The dominant northerly drift of these having been established, rationally or irrationally, at some time in the geologic past, I see no reason for change now. On stratigraphic grounds the pattern seems to have been established by Permian time at latest and has persisted ever since, with especial vigour at the Miocene when India travelled from Australia to its present position, finally ridging up the Himalaya.

As I read the conflict from above the equatorial plane, both streams (northern and southern) were radial, laminar flows originating deep within the mantle body and moving thence out towards the global surface. The flow was extremely slow, taking place largely in the solid state. The two streams, above and below the equatorial plane respectively, were virtually parallel and involved much mixing between bodies of almost identical composition. When the global surface was approached (from below) extra space became available by radial sphericity, and a new dimension 'direction' became significant in place of the prior compression, all-powerful in the depths of the Earth. Gases were now freed in quantity and volcanism was released.

Cretaceous time, when the kimberlites were widely ejected and many samples of ultradeep types of rocks and minerals could have represented this epoch when the southern shove was established and the northerly drift developed in consequence far across the northern hemisphere.

Here are one man's thoughts on how, when Gondwanaland was bursting apart under continental drift, the Laurasian lands were gathered together in laager against the shove from the southern hemisphere.

Was the force a northern pull or a southern shove? In the northwest Pacific Ocean one's first thought might be that it was a northern pull of the sea floor. But the evidence brought from the sedimentology of the region by Heezen and MacGregor (1973) stated clearly that the operation of horizontal displacement *began south* of the equator. This was continued by passage over the equatorial zone, and thence far to the northwest of that great ocean basin; and the story ends with the Gondwana fragments, from Arabia to India. All these had (and have) a stratigraphic sequence of Gondwana (southern) affinity in Palaeozoic time and have subsequently broken away (Fig. 2), crossed the tropical zone and lastly become affiliated to the southern flank of Laurasia, the *northern* continent.

All the transfers to the northern hemisphere have been *real* and, so far, *permanent*. Together they tell of a southern shove which (as with geomagnetics) has caused similar bypolar operations to become *one* global system in actual function.

Chapter 6

Evidence from rock magnetism

Geomagnetic researches and continental drift

When continental drift was first mooted its chief opponents would not accept the idea of continental displacement until they knew 'how it worked'. We still don't know; but later geophysicists provided fresh evidence in favour of continental drift. Briefly, Irving, Creer, and others investigating the properties of magnetically sensitive minerals in rock samples discovered that these properties were oriented incompatibly to the present magnetic field enveloping the Earth.

The explanation was that in igneous rocks the remanent magnetic parameters accord with the geomagnetic field prevailing at the time when the temperature of cooling passed the blocking temperatures of its iron mineral constituents. This magnetism is therefore fixed in direction, or 'frozen' within the host rock; and any apparent change subsequently can only be by tectonic disorientation of the rock mass (which is measurable) or by chemical changes in the rocks.

Certain sedimentary rocks of the 'red sandstone' type may also possess remanent magnetism. Here the explanation is given that when the rock particles are laid down the already magnetized grains tend to orientate themselves in conformity with the prevalent field of magnetic force, which thereafter continues within the sample because the grains are locked in position within the rock. The only apparent disturbance possible thereafter is again by tectonic disorientation or by chemical change.

In the laboratory, to determine the true palaeomagnetic values ('hard magnetism') later magnetic effects have to be eliminated by 'cleaning' the sample by subjecting it to rapidly changing alternating fields on three mutually perpendicular axes, or to heat and afterwards cool the sample in zero magnetic fields.

Specifically, the palaeomagnetic parameters indicated quite different positions for the magnetic poles during the geological past in relation to the sites from which they were determined. Results were consistent and tallied with the various ages of rock systems as derived from palaeontologic and radiometric methods. Either polar wandering or continental drift had occurred; and as the 'polar wandering curves' for the several continents were different the inevitable conclusion was that the continents themselves had drifted not only with respect to the 'palaeo-pole' but also with respect to one another! And when the contemporaneous 'wandering curves' were compared they could be made to agree only when the several continents were reassembled in the figures of Laurasia and Gondwanaland as previously specified by their various stratigraphies (p. 94). Furthermore, the two parent supercontinents were shown to have undergone supercontinental drift relative to the palaeo-pole before their fragmentation in Cretaceous time. Thus was the geological work of du Toit affirming the reality of continental drift confirmed geomagnetically.

One assumption remained — that the magnetic poles had stayed in the vicinity of the geographical poles, for these constituted the ends of the rotational axis of the Earth. This last was demonstrated by palaeoclimatology. The existing major climatic belts of the Earth are defined by the $23\frac{1}{2}°$ inclination of the Earth's axis. These belts are latitudinal. The same distribution is found in stratigraphic geology of the past (e.g. an area of continent which has drifted into the frigid zone has glacially-accumulated rock (tillite), at the same time that another area which is in the hot temperate zone has red, desert sandstones). On the scale of the two supercontinents these effects are displayed very clearly, even contrary climatic effects showing within one large continental mass. When, during the late Palaeozoic, Gondwanaland drifted across the South Pole, for instance, upon the advancing edge in South America warming up was evident when South Africa had just laid down its vast tillite deposits; while upon the trailing edge (India and Australia) which still had to cross the polar zone, chilling towards refrigeration was simultaneously recorded (King, 1962/7).

So continental drift has come to be regarded as 'respectable' and geomagnetic measurements as reliable; but within the latter has developed a sad dichotomy. Whereas palaeomagnetic measurements made on land relate to true magnetic poles of the geologic past and have proved the truth of continental drift; measurements taken at sea and referring (p. 95) to polarity reversals have proved somewhat less reliable than was at first hoped (Vine and Matthews, 1963; Heirtzler *et al.*, 1968)

Geomagnetic polarity reversals at sea

Beneath the oceans, oriented samples for testing are difficult to take. Most geomagnetic measurements are made indirectly by means of magnetometers towed astern of ships or trailed by aeroplanes. These records show the *existing* state of magnetic field along a traverse. At any point the graph records the total field; it does not record how much of this may be 'frozen' hard magnetism, or how much is due to wandering magnetic fields. To monitor these latter changes an American research plane packed with geomagnetic equipment periodically flies around the Earth.

In surveys of magnetism at sea another method is available, involving a recognition of *alternating polarity reversals distributed in stripes on either side of the crestal rifts of sea-floor cymatogens (mid-ocean ridges).*

McKenzie (1970) provided a synopsis of the then current view of polarity reversals: 'The polarity of the main field of the earth is not constant . . . but changes sign causing the North Pole to become the South Pole . . . Though the cause of these field reversals is not known, the reversal time scale for the last 5 million years has been worked out by observing the polarity of dated lava flows.'

'The history of the magnetic field has been recorded by the spreading sea-floor just as a tape recorder records on magnetic tape. Since the magnetization of the basalts on the ocean floor contributes to the field intensity at the sea surface, the evolutionary history and spreading rate of a ridge can be obtained simply by towing a magnetometer behind a survey ship, and matching the records with those calculated from the reversal time scale.'

The magnetic anomaly is very strong at the rifted centre line of the ridge, above the core of rising, mantle-type rock (Fig. 23); but the intensity of the anomalies falls away rapidly to either side, where they may fade away altogether across some ocean trenches or in other specific areas.

But let us be very clear — what the magnetometer has measured is a present total magnetic field. While this may have developed by migration laterally away from a zone of origin, there is no proof that the rocks through which the magnetic impulses now pass have themselves moved laterally with time.

A series of magnetic signals generated at a linear focus could be propagated laterally through fixed rock masses of a stressed lithosphere as sound travels through water. So patterns of polarity reversals may be transient phenomena chasing one another through the suboceanic basalts (and through the continents also) without physical translation of the rocks themselves. Such transitory patterns may be valid, derived in order from the ridge, and datable geomagnetically, without necessarily bearing any direct relationship of origin, age, or function with the host rock for the present time. Nor would they have any necessary relation with any truly remanent magnetism present in that host rock. They may even produce a pattern identical with a record of magnetic 'frozen' reversals.

That there is a pattern of polarity reversals is agreed; that these are 'frozen' into the rocks is *assumed*, and that the pattern demonstrates the physical transportation of rock masses beneath the sea-floor is also an assumption that will be true only if the magnetism is 'frozen' into the sea-floor basalts or other rocks.

Even the outstanding paper by Heirtzler *et al.* (1968) dealing with the pattern of geomagnetic reversal stripes which are parallel, and bilaterally symmetrical, to the mid-ocean ridge system, and which 'may be regarded as strong support for the concept of ocean floor spreading' does not tackle the fundamental problem before us. Though the authors state their assumptions scrupulously and follow the evidence clearly with respect to each of their problems in turn.

Perhaps the authors took for granted the earlier statement by Vine (1966) that the anomaly was caused in the rocks of the oceanic crust 'as they solidified and cooled through the Curie temperature at the crest of an ocean ridge, and subsequently spread away from it at a steady rate.'

The matter needs to be probed further. Actual geographical displacement of crustal rocks has yet to be demonstrated.

The timetable of polarity reversals

The *order* of polarity reversals is determined by comparison of geomagnetic records: the *date* assigned to each needs to be established by correlation with either known fossiliferous sediments or by the radioactive qualities of the rock (K–Ar isotopic ratios is the method commonly used).

To geologists, any chronicle of events relating to the Earth is of interest and the polarity reversals method has two desirable qualities: (a) it is adaptable to the spans of later geological time; and (b) it operates conveniently over oceanic areas where the classical methods of geological study cannot be applied.

Deep-sea cores penetrating to the oceanic basement yield not only a record of sedimentation but, with K–Ar datings of basalts, can be compared for sequence. When a sequence is discovered it usually displays an increase of basalt age outwards from a submarine ridge, which is regarded by Earth scientists as indication that the geomagnetism is truly 'frozen' and the ridge is a 'spreading ridge'. Maxwell (1969) and Maxwell *et al.* (1970) obtained such results and reached such a conclusion for the South Atlantic. But this is not necessarily so. In sedimentary sequences normal overlap up the flank of a ridge (Fig. 29) provides a similar distribution of age-date as the basin filled up.

The authoritative timetable of polarity reversals at sea extending back 80 Ma to reversal 33 is by Heirtzler *et al.* (1968). To this Larson and Pitman (1972) carried the record back to 160 Ma BP (Upper Jurassic) to reversal M*22 (Fig. 24). No oceanic crust is known older than 170 Ma (Fig. 25). Older reversals from the lands dating back to the Tournasian (early Carboniferous) (almost 350 Ma age) have been considered by Irving and Pullaiah (1976); and the former has carried the story of geomagnetics back to the Devonian (Irving, 1980).

The table is founded upon an assumption that the observations giving rise to it are of geomagnetism 'frozen' into Earth materials. But this condition is not necessarily so. Let us take an example.

---

* M for Mesozoic, additional to the younger 33 Cenozoic reversals above.

98

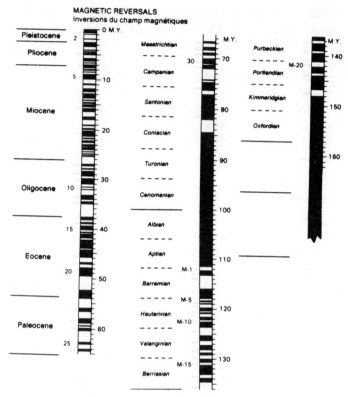

Figure 24   The timetable of geomagnetic polarity reversals at sea. The table from Recent through Cenozoic to Upper Cretaceous was compiled by Heirtzler *et al.* (1968). The Mesozoic record was added by Larson and Pitman (1972). Older reversals from the lands continuing back to the Tournasian (early Carboniferous) have been considered by Irving and Pullaiah (1976)

The South Magnetic Pole

The South Magnetic Pole was first visited on 16 January 1909 by Professor (later Sir) Edgeworth David, Professor (later Sir) Douglas Mawson and Dr A. F. Mackay. Dragging their sledge behind them, the three men trudged over the high plateau of Victoria Land. Their three-month journey, conducted in the Heroic Age of Antarctic exploration, had been of extreme difficulty and hardship.

Some days before, as they approached the Magnetic Pole, they found 'that the polar centre executes a daily round of wanderings about its mean position'. So they calculated 'the approximate mean position of the Magnetic Pole' for the day of their arrival and at 3.30 in the afternoon, as the Pole passed beneath, they bared their heads, raised the Union Jack and took possession of the South Magnetic Pole in the name of King Edward VII. A photograph (Fig. 26) records the little ceremony.

Figure 25   An early isochron map of the ocean basins (Heirtzler *et al.* 1968)

Figure 26 An historic occasion in 1909. A. F. Mackay, Edgeworth David and Douglas Mawson at the South Magnetic Pole. Since that date the Magnetic Pole has shifted 500 km to the northwest; but there has been no corresponding movement of land or sea masses. Clearly, magnetic fields move freely through the rock formations upon a global scale. Magnetic polarities can be captured and fossilized by the iron-rich minerals only where those minerals are momentarily forming (T. W. E. David)

This, and the following note, refers of course to the main dipole field of the Earth, which is perhaps due to reverse itself during the next few decades.

Geomagnetic shadows

Since those days the Magnetic Pole has continued to wander until now, *70 years later, the mean position of the Pole is situated 500 km farther to the northwest*, beyond the very coastline of Antarctica (Fig. 27). This has been a directionally definitive shift of the Magnetic Pole during the lifetime of persons now living.

Of course, if an investigator had set up an observatory during the past 70 years at some point between the 1909 and 1975 positions of the pole he could have found that in addition to the diurnal peripatetics of the pole he had experienced a complete reversal of polarity as the pole (magnetic) changed position from south to north (geographically) of his observatory. In the meantime the position of the geographical pole of rotation would not have changed, nor would the position of the Antarctic continent, nor its rock formations. The only variable would have been the magnetic field passing through the Earth-body as strain shadows pass in polarized light through the fixed molecular structure of a quartz crystal. Both phenomena are transitory and both are polarized. Surely, with a global reversal of polarity all that will happen is that all the stripes formerly showing normality (or conformity with the present dipole orientation) will appear non-conformable with the dipole orientation and vice versa. And there will have been no displacement of mineral fragments or new rock product at all. Change the dipole again and all the stripes change back again. Winking 'on and off' the stripes may well convey a false impression of lateral movement which is not, in fact, there. This is the basis of apparently-moving electric advertisements!

The North Magnetic Pole has also travelled (almost due north) through the Canadian Arctic from latitude 70°N to latitude 76°N, a distance of about 576 km during the time interval 1831–1975 (Fig. 28). Visitors to this pole were Ross in 1831 and Amundsen in 1904.

Travel of the magnetic poles (*and their attendant global fields*) in modern times assures us: (a) that the Earth's magnetic field is independent of the local composition of crustal rocks; and (b) that the Magnetic poles and their attendant fields change in position through all or any crustal rock masses. Wave movements of various kinds are known to pass through solid and liquid materials, why not the magnetic spectrum of electromagnetic waves also?

In the chronology of magnetic polarity reversals, however, are two exceptionally sustained intervals of normal polarity: one in the mid- and late Jurassic (up to 150 Ma), and the second in the middle Cretaceous from Aptian to Coniacian (111 to 85 Ma). These special, non-reversing intervals, which coincide on stratigraphical data with the two main stages of Mesozoic continental drift, are recorded by Scrutton on the Rockall–Faeroes plateau, by Irving and Couillard (1973) on continental data; by Larson and Hilde

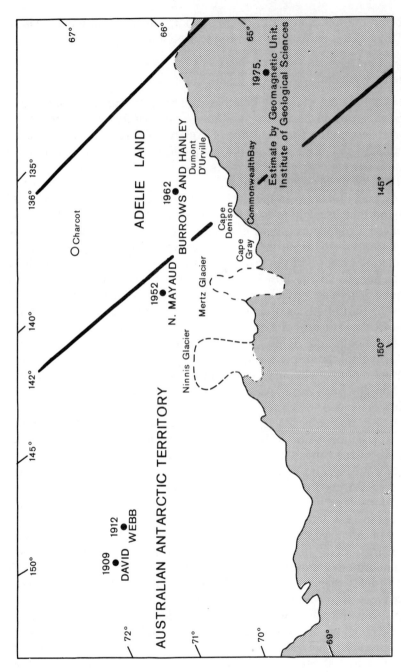

Figure 27 Plot of determined positions of the South Magnetic Pole 1909–75, showing a change of position by 500 km (by courtesy of Dr D. R. Barraclough)

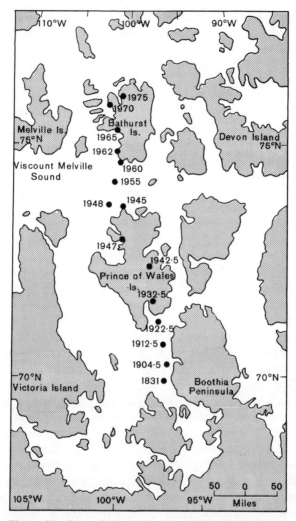

Figure 28   Plot of determined positions of the North
Magnetic Pole between 1831 and 1975, showing travel
northward of approximately 550 km (by courtesy of Dr
D. R. Barraclough)

(1975) from the ocean basin near the Hawaiian Islands. They also coincide
with two phases (early Jurassic and late Cretaceous) of maximum eruption
of plateau basalts upon the continents. All of which points to optimal lique-
faction of the upper mantle (asthenosphere) and particularly the widespread
occurrence of mobile, volatile-charged magmas (with a possible seismic velocity of 6.2–7.2 km/s). These observations may perhaps be indicative of
episodes of low crustal stress around the globe.

Magnetometer traverses, mostly by air, have led to the belief that the field has been weakening for the past 50 years, and that about the year 2000 AD a reversal of polarity may be expected over the entire Earth. These observations prove that the total magnetic field is not fixed in the rocks of the Earth's crust, but that it changes secularly in position. *So, change in distribution of the Earth's magnetic field does not signify change in geographic position of the Earth's crust, unless the magnetic qualities were 'frozen in' (fossilized) at the time the respective rock masses were formed.* Anybody who has crossed the Pacific Ocean has a very clear idea of the magnitude of the problems of crustal translocation there with which a sea-floor spreader must necessarily become involved — by an original dubious assumption (p. 86).

Magnetic reversal patterns in the northwest Pacific are very complex and have been interpreted as showing that the northwest Pacific basin migrated *northward* by 40° of latitude since Cretaceous time; and magnetic anomalies in the marginal seas are quoted as 'even more complex'.

The manner in which polarity reversals are achieved is not certainly known. Two views are held: (a) that the dipole field fades away leaving only a non-dipole field which is replaced by an increasingly reversed field; or (b) that the dipole field 'flips over' and that the dipole character is never lost. This latter view is supported by the transitions studied in core RC14–14 from which Opdyke and his collaborators (1973) were able to trace the 'virtual geomagnetic pole' position for all these transitions. For the first transition (Upper Jaramillo) the apparent magnetic pole, after a leisurely traverse across Arctic America from Alaska to the southern tip of Greenland migrated rapidly thence across eastern America to Mexico, to near the South Island of New Zealand, to Antarctica whereabout it made many excursions. For the second event (Lower Jaramillo) the geomagnetic pole travelled first about the southwest Pacific and then left from the Ross Sea region, across Australia and Japan direct to the Arctic concluding with an excursion far into the North Atlantic before settling in the north. The third transposition (Upper Olduvai) left from near the northern geographical pole across northern Greenland to Central Europe and Greece, before rapidly traversing West Africa and the Atlantic Ocean to near Cape Recife whence it changed course directly to the Antarctic within which it appears to have contained its wanderings for a space. The transpositions from one hemisphere to another are relatively rapid: perhaps only a few hundred years for the north magnetic pole to become the south magnetic pole and vice versa. Truly the virtual geomagnetic poles alternate their normal polar residence spectacularly across the hemispheres.

Additionally, non-dipole fields are known that migrate eastward or westward; and records from Sitka, Alaska station 'indicate an eastward drift during the past sixty years, in contrast to the predominance of a westward direction of drift observed over most areas of the world for the past several hundred years.'

Malin and Saunders (1974) made a specialized study of this non-dipole part of the magnetic field which shows that from 1910 until 1920 it had moved northwestward through the Indian Ocean by 750 miles almost to the Makran coast. By 1928 it was in Oman and in 1930 moved northward along the Caspian Sea. Between 1935 and 1940 it hovered over northern Russia before passing the geographical North Pole about 1960. At present it is situated in the Arctic Archipelago of North America. These wanderings are independent of rock formations over both continental and oceanic areas, and traverse in 50 years about a quarter of the circumference of the globe.

So the magnetic phenomena generated in relation to the mid-ocean ridges, and in particular the twinned polarity reversal stripes recorded by magnetometer traverses across the oceans, may well be only 'signals' the pattern of which passes away to either side from the crestal zone of the ridge *through* the ocean floor, without any necessary displacement of, or addition to, the floor itself. On this viewpoint, sea-floor spreading is an unnecessary and perhaps wrongful assumption. And as we have seen in the Antarctic, it even becomes contrary to established observation of great geomagnetic changes with no corresponding alteration of geography or of rock systems whatever.

The speed of geomagnetic polar change at the present time (about 8 km per year) is also very different from the rates of supposed crustal movement of sea-floor spreading (quoted by geomagnetists as 1 to 8 cm per year).

Also, there is a dearth of reliable information on the distribution of magnetism and magnetically sensitive minerals beneath the sea bed. Dykes may make good records by their narrowness and verticality; but in lava flows and ocean bed sediments geomagnetic units must be distributed horizontally over wide areas. Their succession is vertically disposed and unsuitable to delineation on a horizontal plane (Fig. 29).

Differences occur, moreover, in different rocks or structures and these are often difficult of explanation. For instance, of rocks on the mid-Indian Ridge, Vinogradov *et al.* (1969) recorded differences of intensity in the remanent magnetism which they thought might indicate that 'the magnetic susceptibilities of deep rocks are greater than those of intrusive basaltic lavas. Neverthe-

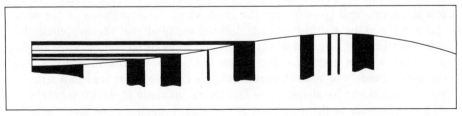

Figure 29   Supposed relationship, by sea-floor spreading, of geomagnetic polarity reversals in oceanic crust (vertical) and in sea-floor sediments and effusions (horizontal) near a growing mid-ocean ridge. A similar picture should result by normal stratigraphic overlap within a basin (either oceanic or continental) adjoining a rising tectonic arch. There could be a similar appearance of successively younger formations towards the ridge — and no sea-floor spreading at all!

less, the polarity of magnetization at the Curie Point may be different at different depths and different times both for deep and surface rocks . . . These differences may be the cause of the stripes of anomalies in the magnetic field which we can observe over the Mid-Ocean Ridge.'

In the same year Bagin *et al.* (1969) found less stable magnetic properties in the magnetite of peridotites and serpentines belonging to the Rift Zone (continental type) where the magnetite was a secondary mineral, than in titanomagnetites of basalts cooled in a magma chamber which showed extreme magnetic stability.

Again, geomagnetic discontinuities are known in all the oceans and these may well be global and occur at comparable intervals of time. If so, they may reflect changes of mantle activity. A comparative survey of results from the southern oceans yields new points of view (Norton and Sclater, 1979; Stevens, 1980).

The oceanic pattern of striped polarity reversals fades away usually as the underwater continental slope is approached. Elsewhere the anomaly stripes 'dive' beneath the continental margins in a manner that lacks a logical or satisfactory explanation.

But the outstanding mystery of paired polarity reversals is: why the reversal patterns should divide and pass away to either side of the crestal rift. That geomagnetic signals might do so is perhaps feasible, that whole rock formations should continue to do so regularly for millions upon millions of years is less credible geologically.

Iceland and the Reykjanes Ridge

One of the most thoroughly researched submarine ridges is the Reykjanes Ridge of the mid-North Atlantic (Talwani *et al.* 1971) which has a very strong bilateral pattern of reversals. The ridge crest is nearly devoid of sediments; but on both flanks, at anomaly 5 (10 Ma) is a scarp beyond which extends a notably augmented thickness of late Cenozoic sediments.

The Reykjanes Ridge is continued northeastward into the Reykjanes area of Iceland where the tectonic features of the submarine ridge become manifest upon the land (p. 00), and the qualities of the hypothesis of sea-floor spreading may be visibly assessed, and the credibility of the hypothesis be established or no. To guide us herein are the Icelandic geologists themselves, who have written upon it in Bjornsson (ed.) (1967). Notable features are: the axial anomaly of the submarine ridge is continued on land in a strip of young Pleistocene basalts, and the first lateral anomaly of similar sign (about 80 km either side of the axial anomaly) is also recognizable on land by strips of youthful volcanism (p. 00). The eastern trace continues into the central graben of Iceland, the western anomaly continues into the youthful volcanic strip of the Snaefellsnes Peninsula. The faulted median rift zone continues right across Iceland to the northern coast. And lastly, Ward (1971) has interpreted the geology of Iceland as being affected by two large transform fault systems, parallel with those described as crossing the Reykjanes Ridge, in the northeastern and southwestern extremities of the island respectively.

Superficially there is correspondence between topographic features and volcanic activity upon the submarine ridge and upon the island respectively. But, according to Einarsson (1967, 1968), whereas the two lateral traces should (on the geomagnetic interpretation of sea-floor spreading) be older than the axial zone, all three strips of volcanics are of the same Pleistocene age! (p. 00).

The Reykjanes Ridge has been quoted as an admirable example of a mid-ocean ridge with a bilaterally symmetrical pattern of magnetic reversals. But is it typical? Over vast areas of the sea bed there is no symmetry in the pattern of anomalies. South of latitude 35°S the Atlantic Ocean pattern is no longer symmetrical (p. 42). In the southwestern Indian Ocean it fails altogether and the sea floor is dominated by mighty fracture zones (e.g. Agulhas and Prince Edward fractures). Within the Pacific Basin the position of the East Pacific Ridge is such that for sea-floor spreading some investigators have written of 'unilateral convection'. To the northeast the ridge disappears for a space under the western flank of North America to emerge once more in the Gulf of Alaska where it meets the Aleutian Trench at right angles. There, according to Beloussov (1970), 'the pattern of anomalies is reversed'. Along the northwestern margin of the Gulf of Alaska, magnetic anomalies can be traced across the Aleutian Trench and for 50 km thereafter into the continental margin; whereas to the northeast the anomalies lose their identities several tens of kilometres before the continental margin is reached. Considerable migration of spreading axes simultaneously in several directions has been postulated here.

An important contribution in this regard is the analysis of crustal stress in the vicinity of Iceland by Hast (1969) in connection with that island and the Reykjanes Ridge. Because of the great importance attached to this area by early proponents of sea-floor spreading he 'performed exhaustive stress measurements . . . the results of (which) are inconsistent with the theories. In Iceland bedrock stresses are compressive and not tensional "even near the Ridge", and the maximum horizontal pressure on the horizontal plane *acts in a direction at 45° to the Mid-Atlantic Ridge and not in the 90° direction to be expected* if new material was extruded to form the Atlantic Ocean floor.'

'The (stress) field varies in magnitude according to the strength of the rock; but everywhere it is a horizontal compressive field and at all depths much higher than the weight of the overburden.' The horizontal principal shear stress in vertical sections through the crust is *parallel* with the sea-floor ridge, and coincides with faults and fractures in the bedrock. These probably continue to great depth, and could provide channels for levitating magmas without a state of crustal tension.

The faults transverse to the ridge are the conjugate shears of the stress field (Hast, 1969), so that *the Reykjanes Ridge is not a tensional ridge, but a faulted ridge of shear*. There is no tectonic reason therefore why geomagnetic anomaly stripes should be paired on opposite sides of the crestal rift as an original and fundamental quantity at all.

One could adduce other examples which have come under discussion during interpretation, but this is not the issue. The basic query is whether the striped pattern of magnetic reversals in the oceanic areas is any more than shadows flitting through the rocks perhaps as a result of crustal stresses (p. 107). As is well known, 'if stress is applied concurrently with some magnetization process it may affect the remanence produced; if applied long after the acquisition of an initial remanence it may modify it' (Irving). The state of stress is closely linked with magnetic remanence, and Irving even notes the occurrence of 'stress-aided viscous remanent magnetizations.' The experiments of Shive (1970) have shown how magnetism, remanent in nickel, magnetite, and other substances, is often accordant with local stress fields. Maybe the magnetic vector migrates into coincidence with a shear-vector (as at Reykjanes Ridge), and this may be so upon many of the world's oceanic ridges.

No vertical shear is apparent in the Icelandic bedrock. Thus theories of rising convection currents are inconsistent with results from stress measurement in Iceland.

As a final comment on polarity reversal patterns, we note that paired geomagnetic anomalies become small approaching an island arc from seawards, possibly because the basaltic crust lies deeper in the slab of a subductogen; and we quote: 'Any semblance of symmetry vanishes most definitely in the case of detailed magnetic measurements with an instrument lowered to great depth, when the magnetic field disintegrates into an enormous number of small ovals arranged in echelon-like fashion.' This has been remarked by many observers.

It appears easier for a geomagnetist to visualize that incalculable quantities of crustal rock should move aside through thousands of kilometres to produce striped patterns of magnetic reversals: it is easier for a geologist to leave the rocks of the Earth in place where he finds them and let the magnetic polarity reversals pulsate laterally through the rock masses and sometimes perhaps align themselves with the shear pattern of the rocks.

Using the light spectrum all petrologists have observed the migration of 'strain shadows' through crystals viewed in polarized light; and (p. 100) two eminent geologists at the South Magnetic Pole (1909) found the dipole magnetic field to travel through the rocks beneath them with hourly ease.

Geomagnetic records in sea-floor sediments

As geomagnetism is controlled by *time*, geomagnetic records within the sea-floor sediments should be *layered* horizontally, and in clear instances geomagnetic reversals should agree with the palaeontologic records of individual strata. Successive records should, of course, be read stratigraphically, i.e. in vertical succession upwards. Many sea-floor sediments, especially those of pelagic type, are however deficient in magnetically sensitive minerals and afford no useful data.

But in the growing libraries of sea-floor drill cores some useful cores are listed. Such a one is core RC14-14 from the Indian Ocean, studied by Opdyke, Kent, and Lowrie (1973), which yields precise vertical information. The core is rapidly deposited 'radiolarian and diatom lutite', and is 26 metres long. The top 460 cm was reversely magnetized; normal polarity is recorded from 460 to 940 cm; after which a long unbroken stretch is reversely magnetized to a depth of 2250 cm where a polarity change again occurs and continues to the bottom of the core. 'Each polarity transition is characterized by a distinct drop of remanent intensity.' Absence of the latest sediments indicates that the core must be older than 690,000 years; and there are other hiatuses in the stratigraphy which, however, other cores show not to have been regional. So, *under favourable conditions*, both natural and operational, satisfactory geomagnetic data of polarity reversals can be garnered from deep-sea drill cores.

In addition to the main table of magnetic reversals, local, short-period reversals are generally noted. These latter fade away with depth and cease to be recognizable in deep-sea cores. Recently cognizance has also been taken of magnetization changes caused by burial and uplift (Pullaiah *et al.*, 1975). By experiment it was found that rocks buried to 5–10 km for several million years acquired a magnetism which should be removed by standard magnetic cleaning procedures. Further experiment at temperatures between 400 and 530 °C yielded a result which indicates that 'very little, if any, remanence can survive high greenschist facies metamorphism, and that the age of the remanence in high-grade metamorphic rocks is between the Rb–Sr whole rock isochron and K–Ar mica ages . . . It is also argued that burial magnetizations are being generated in the present geomagnetic field in rocks at depths of the order of 10 km and are a potent source of magnetic anomalies over the continental crust.'

For assessment of the configuration of the geomagnetic field during Cenozoic times it is necessary to know the actual and relative movements of individual crustal plates (Chapter 7) from non-geomagnetic data. Hailwood (1977) attempted to derive the geomagnetic field from: (a) Heezen's and MacGregor's theory (1973) involving equatorial transit of the West Pacific plate, derived from sedimentological data from deep-sea cores (p. 136); and (b) by reference to the Icelandic 'hot spot'. But each of these tests provided only an inconclusive result.

But, without the cores, how does an aerial geomagnetic traverse recorder sort out the several successive reversals of polarity at the several levels along the line of section?

110

The presentation of geomagnetic data

The presentation of results from geomagnetic reversal studies has its own objections. The first of these is the quite false increase of confidence engendered when passing from the original graphs of the magnetometric traverses (which are not always easy of interpretation) to the final black and white figures drawn as solid Earth-blocks (Fig. 30). Perhaps the reversal anomalies do move outwards from the mid-ocean ridges, and in doing so they would have true chronological significance, but they may travel (as reversals) through an unmoving rigid crust — and there would be no sea-floor spreading. After all, bits and pieces of the mid-oceanic ridges show a measure of local lithological control in the global set up. In the Arctic Ocean there is an apparent change of geomagnetic spreading centres, without corresponding crustal translations. Other examples of movement of spreading centres are known in the other oceans, e.g. in the East Pacific (Herron, 1972), and other instances have been quoted where one spreading zone slowly disappears and is replaced by another. Altogether some correlation may be suggested between geomagnetic fields and the state of stress locally present in the Earth's crust.

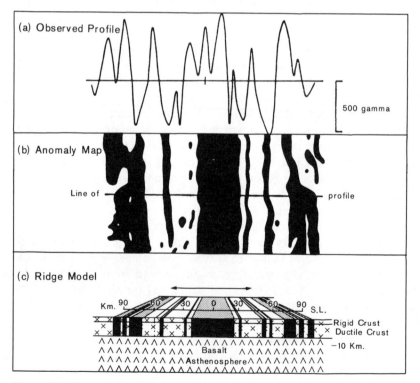

Figure 30 From suggestion to confidence — from magnetometric record to block diagram

The Vine and Matthews (1963) hypothesis which requires that reversal patterns are frozen into the sea-floor rocks is only assumptive. But the reversal blocks, as drawn (Fig. 30) convey a degree of confidence not yet warranted by the original data.

The most dubious case of palaeomagnetism providing a conclusion apparently contrary to the facts of stratigraphic geology comes from Madagascar. According to its exponents, palaeomagnetic data insist that Madagascar, in Gondwanaland, was definitely sited off the eastern horn of Africa. With that as a pre-condition, I have watched a symposium of them (with representatives from England, America, Australia, and South Africa) trying to 'Reunite Gondwanaland'. The reconstruction proved unsatisfactory, especially in West Antarctica and the Scotia arc.

But the original position of Madagascar was alongside Moçambique (Flores, 1970; Kent, 1972; Tarling, 1972) (Fig. 2). The first of these authors was well known to me as the Chief Geologist to Moçambique Gulf Oil, who was also well informed on the geology of Madagascar.

Maybe Fig. 2 of this book offers a better reconstruction of Gondwanaland than the Smith and Hallam diagram used by the geophysicists. For example, the excellent account of evolution of the Indian Ocean basin (Norton and Sclater, 1979) contains the following: 'Between Africa and Antarctica the Prince Edward fracture zone is well mapped near anomaly 34, but due to the possibility that it was a leaky transform [sic] at this time its azimuth may not trace the true direction of relative motion between Africa and Antarctica.'

Chapter 7

The plate tectonics theory

Definition and activity of crustal plates

Following Blackett, the geometry of the plate tectonics method was given by McKenzie and Parker (1967) using Euler's theorem, which states that any motion about the surface of a sphere may be interpreted as a rotation about a pole upon that sphere. These authors postulated that the outer shell of the Earth (50 km or so thick) consists of a relatively small number of rigid crustal plates, some of them in motion, that impinge one upon another. Plates must be the full crustal thickness down to the upper mantle. The relative motions between the edges of the plates induce earthquakes, and the distribution of shallow earthquake foci is then said to define the margins between individual plates (Fig. 23).

The rotation of individual plates may be thought of as 'twizzle' movements about suitable axes within the Earth whose points of emergence are the 'poles' to which reference has been made. Simple vector analysis then shows that theoretically the boundaries of rigid plates may be extensional (ridges and rifts), compressional (trenches), or megashears.

J. T. Wilson (1965) argued that the presence of rifts at the crests of submarine ridges indicated that the central ridge structure was extensional at right angles to the ridge direction. He then derived a theoretical class of faults (transforms) also at right angles to the ridge direction.

Elsewhere (p. 37) we have already preferred Hast's opinion that the rift direction is a direction of shear, and the abundant faults at right angles belong to a conjugate system of shears (Fig. 14). The special class of transforms is then unnecessary. In its place we regard the ridges and domes associated with rift valleys as formed by vertical cymatogenic displacement (p. 38).

As we have noted (p. 38), the measurements by Hast (1969) have shown that a state of horizontal compression is almost universal in the outer earth. (It is, indeed, axiomatic of gravity acting centripetally within a sphere.) Even the mid-Atlantic Ridge is then a shear structure formed under compression, and the crustal rift is only a shallow, secondary structure caused in the superficial rigid crust as a result of the cymatogenic arching by vertical tectonics within the ductile crust and the asthenosphere below, wherein is no tensional effect at all.

The exercise is spoken of as *plate tectonics*, and to begin with *six* major plates (African, American, Antarctic, Eurasian, Indian, and Pacific) were designated. The continents were rafted on the plates and hence they too have moved (drifted) (Fig. 31).

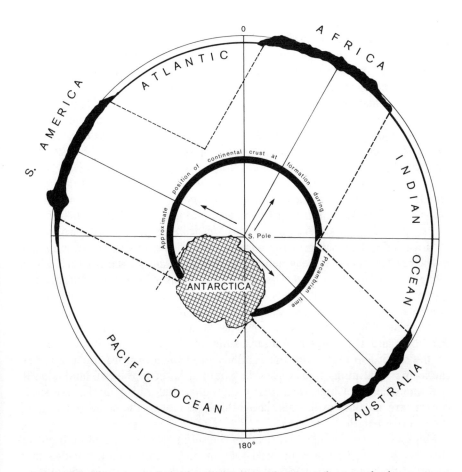

Figure 31 Present southern hemisphere continents and ocean basins on an expanding Earth. Original size of continental crust in heavily-drawn inner circle. Continental drift is shown by vertical rise of the continents on diverging radii, and growing distance between them on the increasing circumference. As areas of continental crust do not enlarge during global expansion, the increase of surface is taken up in the oceanic areas which become wedge shaped. Probable order of ocean basin formation is: (1) Pacific Ocean; (2) Indian Ocean; (3) Atlantic Ocean. South America and Africa are drawn from equatorial section, Australia at 15° S latitude. Antarctica is shown in plan near its present position

With time, sundry modifications were introduced, and by 1976 the original six plates were subdivided to 'at least 12 major ones, and more than 30 minor plates.' Then the heterogeneity of the Earth began to force itself upon the attention of theorists and more complex solutions are nowadays proposed to meet 'special cases'. Interpretations may be expected to become progressively less geometrical and more geological as geological strictures are increasingly applied as criteria for the testing of solutions.

Only at plate boundaries need deformation occur. The interiors of plates are usually stable. The world distribution of shallow-focus earthquakes 1961–67 (Fig. 23), however, shows that beneath the oceans the shocks are clustered along the major sub-oceanic ridges and megashears, which may therefore define the margins of sea-floor plates. Under the continents, however, the shocks are much less localized. Actually the predominant clustering of these shallow-focus shocks is circum-Pacific. They lie along the long-established crogenic belts which (with the Alpine–Indonesian orogenic belt) were in even earlier times parts of the orogenic circumvallations of Laurasia and Gondwana. As Fig. 23 shows, *the distribution of modern shallow earthquake foci is still governed more by the fossil-state orogenic belts of the Palaeozoic, rejuvenated in later times, than it is by the modern cymatogenesis of the sea-floor ridges;. This fact should not be overlooked in assessing modern global tectonics which are active particularly upon sites inherited from a former pattern, now broken up and redistributed about the globe; but still functioning in new global positions assumed in late Mesozoic or subsequent time.*

Megashears (transcurrent movements) occur both on the continents and in the sea floor where movement takes place parallel to the fracture, either straight or smoothly curved. Displacement is mainly horizontal, but in some instances it is also in hundreds of metres vertically (Fig. 14). Examples on land are: the San Andreas and New Zealand Alpine faults; at sea: the Mendocino, Clipperton, and other massive movements of the North Pacific sea floor and the Eltanin fracture zone of the southeast Pacific. In the southwest Indian Ocean are the Agulhas and Prince Edward fracture zones, the former continuing right across the South Atlantic to the Falkland Islands.

Romanche and Chain fractures of the equatorial mid-Atlantic, like many others, are uncomplicated and show displacement in the same sense along their entire length.

Megashears are found with offsets of the submarine ridges and their crestal rifts (p. 36). As in the equatorial section of the mid-Atlantic Ridge they receive adequate geometrical explanation under Euler's theorem; but do they receive adequate *geological* explanation, especially with regard to: (a) the concept of sea-floor spreading with its requirements for new material to be generated at the centre of the oceanic ridges (p. 29), and disposal of old ocean floor at 'subduction zones' (conveyor-belt hypothesis of sea-floor spreading); and (b) does it fit with the hypothesis of *thermal* control of the convection system; or, alternatively, how would it fit a control by de-gassing of sometime mantle material (p. 36)? Finally, (c) how does Euler's theorem fit with the centrifugal dispersion of the southern continents so well established for the late Mesozoic by geological data from those continents (Fig. 31)? Is it necessary to explain these data by plate tectonics at all?

Crustal plates, which are defined with reference to such modern topographic features as mid-ocean ridges and deep-sea trenches, may be no older than those features are (10–20 million years) (p. 175); and oceanic crustal plates would be of very different geological ages from one part to another within a single large modern plate if we are to accept the sea-floor conveyorbelt hypothesis. Useful though it be, Euler's theorem has no built-in time factor that demonstrates continuous continental drift or ocean basin expansion from the Cretaceous to the present day. Arguments currently advanced to explain ancient phenomena are normally based on interpretations from existing topographic conditions and distributions; but these may be oversimple and misleading.

The rules of plate tectonics

The assumptions or rules of the plate tectonics game have been:

(a) Oceanic ridges (e.g. mid-Atlantic Ridge and Antarctic Ridge) can migrate as continents do, in accordance with: (i) sub-crustal forces of the asthenosphere; or (ii) the requirements of Euler's theorem on the behaviour of the rigid crust.

(b) The ridges and trenches may be displaced by the leading edge of a drifting continent (e.g. Andean Trench).

(c) Lithospheric plates can change in size and shape, and can even be destroyed by resorption if the migration of a plate takes it into a trench. Thus the northeast Pacific plate has been almost entirely resorbed at the Aleutian Trench (Pitman and Hayes, 1968).

(d) Ridges (rifted), trenches and megashears do not always appear in their pure form. The 2700 km long, straight Tonga–Kermadec Trench, for instance, acts simultaneously as a trench for the southern portion of the Pacific plate and as a megashear boundary for the Australian plate. It exhibits strike slip. On occasion, ridges may also exhibit strike slip, though this is contrary to the assumption of such sites as extensional.

(e) The Earth is believed to be in a steady state both with regard to density and to surface area. If the hypothesis of sea-floor spreading be accepted, this means that the area of new crust generated along a ridge must be offset by an equal amount of crustal resorption in a trench. Alternatively, some hypothesis of global expansion needs to be postulated (p. 124).

(f) Continents are not capable of being resorbed into the mantle, owing to their buoyancy and hyperfusible petrologic make-up.

(g) New crust is always basaltic and is greatly charged with volatiles. It may be injected along zones where plates have moved apart or where megashears are irregular.

(h) Crustal plates move in response to forces acting on them, either laterally or from below. Movement may be in any direction around the surface of the globe. *The applicability of Euler's theorem is a property of the Earth's sphericity. It is not a function of the geological nature of the applied forces.*

(i) When lower mantle plumes, registered at the surface as 'hot spots', are combined with crustal plate motions, they may account for lines of volcanoes (Tuamotu Ils, Emperor seamounts). Conversely, hot spots may migrate in the mobile mantle beneath a static crust.

(j) Plate tectonics theory does not specify continuity or regularity of action. Intermittent activity is equally admissible, and in this the theory comes much closer to geological evidence than does the sea-floor spreading concept.

The Vectors of differential movement then indicate 'a history of episodic spreading directly related to the major orogenic phases.' Not, we note, to a history of continuous sea-floor spreading. This statement by le Pichon is 'spot on', in accordance with late Mesozoic and Cenozoic geologic history.

(k) The Eulerian axes of rotation for lithospheric plates are deemed to be capable of change, and hence the lateral movement of plates may be interpreted with great freedom. In practice, this frequently leads to conflict with other plates or with geological observations and the freedom claimed is like other 'freedoms' — it is sometimes found to carry a greater measure of responsibility than the originators bargained for.

(l) Plate tectonics theory does not specify any particular energy source for the motion of plates. Some researchers have remained with the thermally controlled, diapiric injection of new basalt at the crest of a 'spreading ridge'; but this is not adequate to the complexities of geological structures and histories of many regions, e.g. the Mediterranean. Three possible motivations for plate activity are generally visualized:

(i) by viscous coupling with a mobile asthenosphere below;
(ii) by pull from localized subduction drawing oceanic crust towards it (Elsasser);
(iii) for small plates, by reaction to larger plates moving alongside.

But as yet there is no final simple solution to the problem of the mechanism of plate tectonics (but see p. 123).

While fracture and shallow focus earthquakes are typical in the thin, upper crust of which the plates are composed, deformation of the underlying asthenosphere or upper mantle is probably by 'creep', as is believed to occur (without observable earthquake activity) along the San Andreas fault below a depth of 15 km.

Remembering that the zone of basaltic crust is worldwide, beneath continents and oceans alike, much more complex and changing convections can be visualized beneath it than sea-floor spreading offers. So the rotational movements indicated for drifting continents by their geologies, for instance, can be interpreted as due to 'whirlpool' or 'vortex' currents in the mantle.

Of these, both North and South America, Eurasia, and Antarctica are indicated as having rotated clockwise. Australia, in its later motions at least, moved counter-clockwise.

(m) Along the sub-oceanic ridges earthquake foci are shallow (10–20 km); but beneath trenches and mountain systems they occur at all depths down to 700 km. Plate tectonics has regarded these zones as extensional and compressional respectively, but this is by no means necessarily so, and a much wider range of phenomena can be introduced by shearing, with both tensional and compressional components of local occurrence. The rest of the Earth surface is relatively aseismic, virtually by plate definition.

(n) Where plate boundaries meet they commonly form 'triple junctions', recognition of which often supplies information on the relevant crustal plate boundaries, collisions and rotations.

(o) Recent mapping of the Falklands–Agulhas and Prince Edward fracture zones (Fig. 2(a)) and their continuations northward through the basins of the western Indian Ocean, draws attention to the development of certain plate boundaries that are partly transoceanic and partly marginal continental. These two fractures first functioned at the break-up of Gondwana during the Mesozoic. During the Cenozoic they were broken into several minor plates which behaved thereafter individually (Norton and Sclater, 1979).

Plate tectonics and orogenesis

Le Pichon (1968) concluded that whether trenches or mountains are formed at leading plate edges is a function of the rate at which the plates are moving together. With plate motions less than 5–6 cm a year the crust can absorb the compression and buckles up as large mountain ranges within which folding and overthrusting deform the crust. With more rapid movement the plate breaks free and sinks into the mantle, creating an oceanic trench.

But later orogenic studies, notably of island arcs and of the Mediterranean region, do not support such a simple explanation. Moreover (p. 188), planation surfaces upon high lands often show them to owe their present elevations not to any original orogenic compression but to repeated vertical uplifts of former denudational plains (p. 187).

From an analysis of sediments, volcanics, structural, and metamorphic chronology in mountain belts, Dewey and Bird (1970) have strongly indicated that orogenic belts of Precambrian to Pliocene age are a consequence of plate-margin evolution. As in the Andes, the andesitic lavas which are typical contain more silica than the ocean-floor emissions and it has been assumed that they are formed by partial melting of the descending slab of a subductogen at a depth of about 150 km. At surface, this forms the inner, volcanic arc of an island arc structure. On land it is further contaminated by granite as the magma rises through continental crust, and the volcanoes may crown high mountain ranges, e.g. Popocatepetl in Mexico, Aconcagua (6960 m) on the Andean border between Chile and Argentine, or the double cones of Elbruz in the Caucasus.

'Alpine peridotites' occur in linear belts, where they display tectonic fabrics. They are high in magnesia, and chrome spinel is common. Most workers consider them to be derived from the mantle. They occur in the Precambrian and all younger mountain systems where they are discordant to the surrounding rocks, and show a high degree of serpentinization (p. 46) with the usual serpentine tectonics developed during later episodes of tectonism.

The presence of such an orogenic ophiolite is nowadays taken as evidence of closing of an ocean between two colliding plates. Such relations are beautifully illustrated by the Cenozoic emplacement of a large ophiolite body in New Guinea, resulting from the collision of the Australian continental mass with the Banda arc (p. 13).

Subduction may, apparently, start anywhere at plate margins where inverse gravitational relations have been produced between crust and mantle. As the heavy oceanic-type slab goes down to be consumed, it is replaced by lighter gas-charged mantle (e.g. Tonga Trench). The site of subduction may then wander (or the plates move) until a passively riding continent reaches the site. This cannot be consumed: it is too light and hyperfusible. Thenceforward, so long as the subduction zone is active, the continental margin will govern the proximity of the subduction zone, and if the margin is advancing it may accrete to the continent a number of successive subduction zones and orogenic belts, e.g. western side of the Americas, both North and South.

The plate tectonics theory was enunciated in the first flush of suboceanic data from remote sensing, as expressed in le Pichon's great synthesis (1968). It offered a splendid new comprehension of global tectonics covering both small and large examples, and rendered new insights into orogenic evolution. Concise generalizations could be formulated, e.g. convergence of the North and South American plates, combined with northeast movement of the East Pacific plate into the Caribbean area, caused the Laramide orogeny (late Cretaceous and early Cenozoic) about the Caribbean.

Wheels within wheels, multiplying the number of plates and tectonic poles, has brought the solutions ever more and more into conformity with local geological complexities of the Earth; and the experience gained enabled the method to be carried back to, at least, Palaeozoic stratigraphies and structures.

Specialized applications became possible and at the time of writing the method is being used in the interpretation of orogenic mineral deposits, e.g. Mitchell and Garson (1976) from whom we quote: 'The relation between these deposits and evolving plate boundaries is reviewed, and the hypothesis is shown to be useful in explaining the origin of the host rocks of ore bodies and hence of the ore bodies themselves.'

'Changes in mineralization through geologic time are related partly to changes in tectonics and magmatic processes but there is some evidence that many types of mineralization have changed relatively little in the past 3000 m.y.'

Already the rotational type of crustal movements envisaged under plate tectonics has great advantages over sea-floor spreading in that the necessary motions are much reduced, e.g. South Atlantic (Fig. 38). There is economy of movement and economy of space. One does not need to think of conveyor-belt tectonics but of crustal plate rotations; and while, in a way, the poles of plate rotation are hypothetical abstractions, several movements with different poles can be combined to achieve coherent results. (For instance, Wellman's (1973) analysis of the New Zealand Alpine fault, segments of which may be referred to individual poles arranged in line, thus explaining sundry irregularities and changes of direction along that megashear. Incidentally, Wellman concluded 'that most of the 450 km displacement on the Alpine Fault Zone has taken place during the last 20 m.y.'; but there was no definite age for the commencement of the 450 km displacement.)

Some results from *Glomar Challenger*

So abundant have been the results from researches aboard the *Glomar Challenger* that they still await the attention of synthesizers; but the following statements culled from later reports of the research supervisors indicate how rapidly opinions are developing on aspects of marine tectonics with especial reference to plate tectonics and the importance of vertical displacements:

'The movement of the crustal plates is far more complex than was believed. They have stopped, started, speeded up, slowed down, and even changed direction, altering the shape of both oceans and continents in the course of their wanderings.' This is in accordance with the experience of geologists.

'The surface of the earth moves up and down vertically almost as fast as it travels laterally. A fragment of continent larger than Great Britain, on which dinosaurs, mammals and birds once lived, has sunk more than 1500 metres in the Atlantic between Ireland and Iceland' (Rockall).

A 'wedge of volcanic ash extending 1500 miles into the Pacific east of Japan took 5 million years to form. Indeed the whole western Pacific bottom is covered with lava spewed up from thousands of undersea vents.' (This is not the timing of sea-floor spreading.)

'The gigantic fan of sediments washed down from the Himalayas by the Ganges and Brahmaputra Rivers extends 2000 nautical miles into the Bay of Bengal, and the force of the Amazon and the submarine river issuing from its mouth has strewn minute topazes, tourmalines and other gems from the Andes as far as the floor of the mid Atlantic.' (In the opposite direction from widening of the Atlantic.)

'All the continents are drifting northwards except Antarctica.' (Is this an orientational illusion? Antarctica is the only one that *can't* drift in any direction other than north!)

Some queries and some doubts

To begin with, plate tectonics was based on the results of remote sensing at sea. Later it was applied successfully to the orogenies of certain continental margins. Within the continents it has had still less application, and continental masses — having had early experience of plate tectonics — seem not to repeat that performance indefinitely but are restricted thereafter by their buoyancy and hyperfusibility to vertical displacements.

Despite the triumphs of the past decade, there remains a need for further checking of data, and for geological nuance that comes only from long acquaintance with this Earth of infinite variety. Cox (1973) expresses this more fully: 'The paradigm of Plate Tectonics, although immensely stimulating to the geologist, has not provided a universal framework that accommodates all tectonic problems. It is a common experience, in fact, for the observations of a field geologist mapping a given quadrangle on a continent to be so dissimilar to what the marine geologist tells him about the plate tectonics of the adjacent sea floor that the two scientists might well be mapping on different planets. Generalizations drawn from plate tectonics are commonly more useful on a global scale than on a detailed local scale. Also the continents are thicker than the oceanic crust and more heterogeneous, reflecting the longer and more complex histories of the continents. As a result, recognizing many of the implications of plate tectonics has been a slow process.'

We have noted that sea-floor spreading and plate tectonics became popular concepts immediately following the acceptance of continental drift, which was already proved by geological data. But, following du Toit, geologists had been careful to relate continental drift to late Mesozoic tectonic activity, which was episodic. The neotectonicists disregarded this point and thought of plate tectonics as a general and continuous process of lateral change. They postulated *average* rates of horizontal movement in the several oceans — averaged over the past 100 million years. In geology, time is long and tectonic averages mean little. Tectonic happenings (both vertical and horizontal) are episodic and not infrequently of global extent, with long quiet intermissions during which wide planations developed upon the lands and ample depositions took place within the oceanic basins.

Stratigraphic evidence and denudational histories confirm (p. 188) that vertical and horizontal displacements have often occurred simultaneously through Mesozoic to Recent time. And anticipating a little the conclusions of a later chapter (p. 187), we can say that there were two phases of continental disruption and drift in the mid-Jurassic and late Cretaceous respectively, followed during most of the early Cenozoic by tectonic quiescence and widespread planation (pp. 189–190) denudational upon the lands, sedimentational in the oceans. Only the India — Australia — Antarctica land mass drifted extensively during the early Cenozoic.

Tectonic activity resumed on the planetary scale (with only local drift) during the late Oligocene to early Miocene, and has increased (with quiescent intervals of widespread synchroneity) until the Pleistocene at least.

The lithospheric thickness beneath the oceans is quoted at 70–100 km. The upper part of this material is currently rigid rock with a seismic velocity of 5 to 6 km/s. How does a plate of such material transmit lateral stress for long periods across the great widths of the ocean basins? If, as is generally supposed, it is isostatically supported, against gravity, upon an asthenospheric upper mantle, lateral transport should be by activity within that upper mantle. Accommodation within the crust should then be by shear, as the Icelandic geologists, who live upon the mid-Atlantic Ridge, have pointed out.

Gilluly (1971) expressed the opinion of many colleagues when he wrote: 'So far as I know, no-one has yet suggested a model for the generation of plate motion that is acceptable to anyone else.' Yet he accepted 'that the reality of plate tectonics seems about as well demonstrated as anything ever is in geology.' In other words, plate tectonics has a better geometry than continental drift had, but as yet, no better motivation. Plate tectonics motions still lack a dynamic theory.

'Certain kinds of tests that have been accepted as proof of the validity of the plate tectonics concept in fact yield only permissive rather than conclusive information. The Joides drill holes, for example, while furnishing invaluable data on the nature of the sedimentary section, have in no case penetrated deeply enough into the basaltic "basement" to demonstrate that it is indeed *basement*, and does not conceal sedimentary rocks below.' (Maxwell, 1973). One has only to think of the thickness of Mesozoic plateau basalts on the southern lands, and of the sedimentary sequences which they conceal, to realize the force of this *caveat*.

Plate tectonics theory has ignored the fact that the orogenic mountain structures of the Earth mostly derive from the circumvallations of Laurasia and Gondwana which were developed during the Palaeozoic or even earlier, and which were actively producing fresh fold-mountain structures as late as the end-Oligocene (Alps, Himalaya) and great cymatogenic arches during the Pliocene and Pleistocene (Andes, Southern Alps of New Zealand).

Dewey (1970, 1974), who made brilliant syntheses of orogenic belts using the plate tectonics method, has nonetheless remarked that 'plate tectonics has given us fresh insights into the kinds of questions to which we may never know the answers, and made us aware of the bewildering array of complexities inherent in orogenic evolution.'

Doubts and uncertainties regarding sea-floor spreading and geomagnetic reversals notwithstanding, most geologists feel that plate tectonics gives more satisfactory answers to the geometry of tectonics than have been available hitherto. And although the ultimate dynamics of horizontal crustal movement have not yet been probed, progress in visualizing global tectonics has certainly been made.

The proposition was summed up by Dietz (1977): 'Although plate tectonics offers a great clarifying principle, in some ways it increases the difficulty of finding a unique solution to any geotectonic problem because it adds a new dimension — in fact a fourth dimension. Formerly, fixistic geology was concerned with up, down and time, but now we must add sideways! For example, the Bahama Platform since its birth, presumably in the Triassic, has undergone a remarkable subsidence of about 6 km; but during the same interval we now learn that it has drifted probably about 6000 km. The evidence for subsidence is derived, of course, from the classical stratigraphy of layered sedimentary rocks, but the quantitive evidence for drift comes from the new stratigraphy provided by the magnetic-anomaly stripes.'

Later in the same work he opines: 'But to accept continuous and steady shifting of crustal plates of but a few centimetres per year comes close to defying imagination.' Stop-and-go drift movements might be more in accord with natural process.

To return to the lava pool at Nyamlagira (Fig. 17) observed in 1939, the congealed skin of lava was seen to sink into the pool at some point or line as though it was being sucked down, but more probably in response to the increasing density of crust as its volatiles were lost. As this slab descended, it drew the whole skin after it and fresh liquid lava rose in the rear to make new surface which congealed in its turn giving new skin. Each time the governing mechanism was the sinking of solid, densified crust, which dragged away the crust behind. Such drift of skin was not uniform but intermittent. The intervals of cessation were much longer than the phases of activity. This mini-example is in accord with the views of Richter (1977) on the probable mechanisms of plate tectonics. The activating part of the mechanism would appear to be the slab of de-gassed, solid crust descending back into the asthenosphere below, and dragging large areas of oceanic crust behind it.

A master mind in geotectonics

Twenty years before the words 'plate tectonics' meant more than an unholy crash offstage in the kitchen, the geography of the globe was analysed by a Gondwanalander with very sound ideas on tectonics in general, and global tectonics in particular.

Beginning to publish around 1938, before the form and structure of the ocean floor were really known, he built his own tectonic system with its vocabulary of oroclines, rhombochasms and sphenochasms, orotaths and nemataths, etc. On these topics he lectured in many parts of the globe.

He held his own symposium on continental drift in 1956, two years before the august Royal Society of London handed the topic over to physicists, and ignored geologists — who had already proved it (A. L. du Toit, 1927, 1937).

He retired at the end of 1976, leaving a valedictory volume, *The Expanding Earth*, by Professor S. W. Carey, University of Tasmania, worthy of the attention of a new generation of geotectonicists.

This is no place to review Carey's book, it must be read; but the hypothesis of an expanding Earth which he advocated can now be examined with respect to the data before us. The familiar world has latitude and longitude wherewith to measure distances and areas upon its curved surface. Both these quantities are dependent upon the radius of the globe, and any general change of radius must result in an increase or decrease of surface by a proportional amount. Mathematical friends tell me that, for a small increase of radius, the increase of surface area is expressed in the formula: $8\pi r$ times the increase of radius, where r stands for the radius.

On the present Earth (radius 6371 km) this means that an increase of 1 m on the radius would provide an increase of surface area about 160 km²; and an increase of 1 km would produce an additional 160,039 km² surface area. The present surface of the globe ($4\pi r^2$) is 570,101,310 km², so the proportions of the calculation seem reasonable for the expansion postulated for continental drift in Phanerozoic time.

Following Wegener and du Toit, indications that continents had drifted apart by lateral movement at the surface of the globe were often discussed but no mechanism was adduced other than Holmes' subcrustal convection currents. Under the hypothesis of sea-floor spreading these were developed into convective 'conveyor-belts' originating along the mid-ocean 'spreading' ridges.

Plate tectonics was seemingly no better equipped with a *modus operandi*. Yet most of the requisite knowledge was already in the possession of geologists:

(1) Volcanic action is, or has been, widespread in the regions where phenomena of lateral movement are apparent. This action is usually phreatic. We have already (p. 112) drawn attention to the power of gas emission at all stages of the volcanic cycle. Fundamentally, of course, emission of gas accompanies the rise of liquid lava and ejection of solid materials, ash, scoria, foreign rocks, including some from the mantle beneath the crust.

(2) The uprush of ejectamenta is vertically directed (*radial* within the Earth). It takes the shortest route from the interior towards the surface, where it rapidly cools in a normal temperature and pressure, losing activity and mobility as it does so. But it has burst through to a level of relatively low energy production, *and increased space*, the surface of the sphere. Moreover, the energies here lack lateral bounds. The surface of the sphere therefore has no lateral confinement, and movements or displacements are often expressed horizontally. Continental drift, or plate tectonics, become an expression of this, with doming and rifting as natural topographic expression.

(3) There is general agreement among experimenters that when the continental masses are refitted on a globe, they fit best upon a smaller globe. This argues that the globe has been expanding at the maximum of lateral continental drift. Dearnley (1965–6) deduced an expanding Earth from a reconstruction of Precambrian orogenic belts.

(4) McElhinny and Brock (1975) have demonstrated that no increase in size of continents has occurred under plate tectonics and continental drift.

These are old masses of volcanic slag and other rocks that are now petrologically dead. Thus the Earth expands, but not the continents. The graticule of surface measurements expands, but the apparent expansion (new space) is expressed only in the oceanic areas.

(5) The major oceanic areas have definitely enlarged, with changes of shape. Beneath the ocean floors, too, this same vertical force of levity is shown in the sheared regions of the so-called 'spreading' ridges.

On the sea floor, areas of vertical uplift or of subsidence (or of lesser vertical uplift) are sometimes signified by the distribution and thickness of relatively young sediments.

(6) With time, too, there has also been a general decrease of gravity which accords with expansion of the mantle. Jordan worked on it.

(7) So there is little wonder that all the southern continents show extraordinary expanses of 'plateau' basalt of an age corresponding with the break-up of Gondwana and the centrifugal flight of its pieces when continental drift was probably at a maximum.

(8) Instead of thinking of lateral (or horizontal) movements on the surface of a sphere, plate tectonicists or continental drifters need to think of vertical (centrifugal) levity *within* the sphere. The higher that the basalt lava columns arise the greater becomes the *apparent* lateral force which is summarized under the term 'an expanding sphere'.

Plate tectonics acting superficially upon a sphere does not lack a built-in cause for a surface expression of *apparently* lateral displacements. Amid natural, universal, *radial* forces of both levity and gravity, aided by a degassing mantle and occasional floods of plateau basalt, there are many possibilities of motive power.

'Upon a sphere'. Therein lies the answer (Fig. 31). The geometry of plate tectonics, with generation of 'new surface' area, demonstrated by increasing separation between continents as the geological history of the world proceeds, is normal and requires that the Earth-sphere shall have expanded beneath the oceans. Plate tectonics not only requires these things but affords measurements of them as the lands pass across the enlarging graticule of meridians and parallels.

What makes it all so clear is the fixed size of the continents, moving (apparently) among so many other variables and leading to an Einsteinian solution that there is change of the graticule measurements of latitude and longitude through which the continents move as the Earth expands.

As Carey sums up (1976, p. 239): 'Outgassing on an expanding earth differs from the convection model in that rising diapiric zones may or may not be accompanied by a returning, sinking zone. *All active motion could be radially up.*'

This subject will be further discussed on p. 175.

Chapter 8

Sedimentation in the world oceans

Oceanic sediments

The oceanic area is approximately 2.5 times the combined land areas of the globe, with an average depth of 3795 m. The sediments that have accumulated in the ocean basins preserve a record of much that has happened on the globe during later geologic time. The ocean bottom is not just monotonous sediment: the composition and grain size distribution of marine sediments change rapidly, even far from land; and there are well marked relationships between marine sedimentation, bottom relief, and tectonics. The sediments are, moreover, commonly zoned in accordance with geographical, climatic, oceanographic, and circumcontinental factors. This information has become increasingly available only since 1950. Yet the deep-ocean-floor plains together make up half the area of the globe.

According to Lisitzin (1972): 'The bulk of (oceanic) suspensates is fine grained. Therefore sorption and desorption in suspension are of especially great significance. The particles suspended under each square metre of ocean surface, ignoring colloids, present more than 30,000 sq metres of surface area; if the colloidal part is taken into account this potential sorbent area is increased by an order of magnitude.' Sandy deposits in the basins or on the continental shelf usually reflect terrestrial weathering or conditions of transport.

The basement on which these sediments rest is the oceanic crust of basalt, dolerite, metabasalt, and gabbro of 6.7 to 6.9 km/s seismic velocity, increasing downwards.

Exceptionally thick bodies of sediment accumulate in areas where progressive sinking of the sea floor adjoins regions of uplift where rapid denudation supplies large amounts of terrigenous waste. This is so about the continental margins, and along volcanic island arcs. The earliest, and most deeply buried layers of sediment come under great pressures and tend to have the water content compressed out of them. As this water migrates upward through the covering layers, the deep layers dry out and develop a light root at the base of the sedimentary pile. As this tends to rise (levitate), and produce deformation (folding and faulting) within the overlying mass of sediments, metamorphism may appear with formation of new and characteristic minerals.

If the original site of deposition was a subduction zone, excess of sediment may clog up the subductive mechanism, and elevation of adjacent lands may be strongly orogenic. This is, for example, the sequence of events leading to the Cretaceous Rangitata Orogeny of New Zealand.

Nowadays the deep-sea drilling programme provides cores that have penetrated for hundreds and even thousands of metres below the ocean bottom. Areas of Cenozoic outcrop may be delineated, and underlying Mesozoic formations can occasionally be sampled. These cores sometimes preserve almost continuous local histories, often undisturbed; but as Moore (1972) points out, even though the tough deep-sea cherts can now be penetrated by the new roller-cone bits, we cannot recover the underlying sections. Upon hitting semi-lithified sediment, or alternating lithified and soft sediment the sampling programme becomes literally a wash-out owing to 'the water which must be circulated to keep the bit clear and the string free'. Technical problems are encountered even in recovering samples of ordinary soft, pelagic sediments or Tertiary muds and 'the recovery of an obviously undisturbed core is sufficiently rare to be labelled a fluke.' Sediments may record local deformation, and by careful interpretation they yield estimates of past changes of depth and changes of climate.

Modern acoustic reflection techniques extend the information from single borehole cores recognizably to long distances yielding distribution data often of great value. Also, in almost every section examined, the ocean-floor sediments reveal vertical faulting and displacement at intervals (Fig. 32).

Special interest attaches to deposits around the continental margins where sediments are thickest (up to 15 km), and from which may be deduced geographical and tectonic regimes that formerly existed upon the adjacent lands (p. 207). Indeed the bulk of sediment contributed to the oceans is supplied to the continental rise as the dissolved, suspended, and colloidal

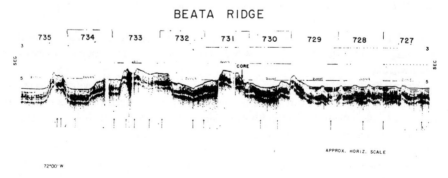

BEATA RIDGE

Figure 32   Reflection sampler record of bottom sediments across the Beata Ridge in the Caribbean. Note the importance of vertical faulting (after M. Ewing *et al.*)

load of rivers, although this may be equalled in polar regions by the contributions of glaciers and icebergs.

Powerful turbidity currents, wherein sediment-laden waters from flooded rivers do not mix with sea water but, preserving their identity, carry their clouds of debris along the sea bottom (where they are powerful eroding agents), sometimes bring coarse terrigenous material and scour out deep, steep-sided submarine canyons. In transit they may spill over the sea floors where they leave characteristic deposits of rapidly alternating, graded coarse and fine layers called turbidites. These are distributed either as great cones off the mouths of major rivers (Amazon, Indus, Ganges), or become ponded as abyssal plains where they bury the bathymetric detail. Turbidity current systems, like those of rivers, converge and join at accordant levels. The thalweg is continuously downstream. Systems are known, e.g. in the North Atlantic, which reach the deepest parts of the ocean basins far from the lands of their origin.

Not all turbidity flows within the oceans originate upon the lands and travel via submarine canyons to the deep ocean floors. Many originate as debris slides upon the continental slope, creating sea-slides (of jumbled structure) upon some of the large earlier-deposited detritus cones.

Between Cape Town and Kerguelen cold bottom currents are deemed responsible for much giant rippling of sediments which normally are horizontal, Ewing *et al.* (1969) write: 'These sediments are neither pelagically draped over the basement topography, nor are they ponded in depressions.'

Amid the turbidites, volcanic ash often makes good marker horizons, particularly in sequences of the Pacific Ocean where they are associated with much finer pelagic clays.

In regions reached by turbidity currents the bottom sediments often show a contrast between pelitic suspensoids (which may be in excess of 90% in some localities) and well-bedded turbidites showing several assorted grades of particles. On the Antarctic shelf, where sediments are sorted by currents and waves, size profiles of suspensates and bottom sediments from the same localities often differ greatly.

Pelagic sediments

In the great ocean basins, and particularly in the Pacific where watersheds are situated not far inland and large rivers are few, terrigenous waste may be available only in relatively small measure. The bulk of sea-floor sediment is then pelagic, derived as dust from the atmosphere and organic remains from the sea itself. Quite different controls govern the laying down of related sediments.

On the world scale, Lisitzin (1972) finds: 'A close relationship between climatic zonality and sedimentation is manifested at the most early stage of sediment formation, even in suspension. Both the quantity of suspended material, and its material composition, depend on climatic zonality . . .

Maximum amounts of suspensions are contributed to the seas and oceans from land where the intensity of erosion is greatest. This can take place only in the regions of maximum moisture, the humid zones (two temperate, one equatorial). These regions coincide with regions where deep oceanic waters rich in nutrients ascend to the 200 metre-thick surface photic zone.' Hence these sediments are often richly fossiliferous. In arid zones, contribution to the sea is very low.

Also: 'great depth is common in the ocean and in some marine basins also influences sedimentation rates and composition . . . The longer a particle resides in the water column the farther the processes of organic matter decay and the dissolution of unstable compositions will proceed. Vertical zonality is particularly influential in the distribution of carbonate sediments.' Below the depth of carbonate compensation carbonate disappears altogether by solution and so is lost from the sedimentary record which becomes silica rich in consequence. This is particularly effective in equatorial waters which are replete with small animal and plant species, and in nutrient-rich zones of deep ocean upwelling along the western shores of continents. In shallow-water deposits all the original organic constituents will be represented in the sea-bottom strata; in deep water all tiny calcareous skeletons are lost. Thus equatorial Pacific sites are typified by DSDP cores 77 through 84 consisting of alternating layers of more or less siliceous calcareous ooze grading downwards into chalks. Many basal sediments here are baked, indicating that basalt is intrusive below.

The ultimate product of suspensates in very deep areas is, of course, the red abyssal clay first made famous by the *Challenger* reports of a century ago. Pelagic red clay from the northwest Pacific has been dated as 145 to 155 million years old, corresponding to early Cretaceous and Jurassic age. Only rarely are sediments older than this encountered in the modern oceans, and then not beyond Triassic.

Except in the region of the West Wind Drift, surface currents of the ocean are generally underlain by deep counter-currents and these may be succeeded by a special deep-sea system of currents. Bottom sedimentation often bears no relation to surface current directions. Tidal currents, by constant repetition, may also be important in some places. Sedimentation is thus controlled chiefly by water dynamics; but the irregular topography of the ocean bottom (which is as dissected and varied as that of the lands) also greatly influences the local distribution of oceanic sediments.

Pelagic sediments derived from the oceans themselves, blanket all submarine topographic features alike. Such sediments, remarkable for their high proportion of siliceous organisms (diatoms) are abundant to the south in the cold waters beyond the Antarctic Convergence.

Certainly one of the most remarkable of the deep-sea drillings (192) was sited on the northernmost of the Emperor Seamount chain off Kamchatka. The top of this isolated massif yielded the surprising succession: 0–140 m Holocene to Pliocene diatomaceous silty clay, diatom ooze with abundant volcanic ash beds and glacial erratics; 140–540 m upper Miocene diatom ooze; 540–910 m middle Miocene claystone; 910–1044 m Oligocene (?) to lower Maastrichtian chalk with calcareous claystone with silts and sands; 1044–1057 m pyroxene plagioclase diabasic basalt. 'Perhaps the most surprising is the sedimentation rate of the middle Miocene claystone, about two orders of magnitude higher than typical pelagic clay rates. This implies proximity to a land area of rapid denudation.'

The oceans have not as yet revealed any important stratigraphic stage not already known on land. The thickness of stage sequences and the hiatuses between them have not revealed any type of sequence that could be called 'unexpected', although there are certain non-sequences (due to local tectonic movements) that evoke a mild surprise. On the whole the succession of beds is what might have been anticipated.

A rough computation (by volume) of the proportions of post-Jurassic sediments in the present oceans is:

Quaternary	1%	
Pliocene	5%	following second uplift of the lands
Miocene	14%	following major uplifts both continental and suboceanic
Oligocene	7%	end of early Cenozoic planation on the lands. Marine Oligocene is often missing over large equatorial areas. Unusually complete Oligocene is found in the Tasman and Coral Seas which were then being formed
Eocene	12%	
Paleocene	7%	
late Cretaceous	24%	following Cenomanian uplift of continents
early and mid-Cretaceous	30%	following disruption of Gondwana and Laurasia

The oldest sediments cropping out within the ocean basins are early Cretaceous. Jurassic and upper Triassic samples are known only in borehole cores — from the northwest Pacific and the northwest Atlantic. Late Cretaceous sediments have been cored from widely dispersed localities in all the oceans; but principally about the continental margins. Paleocene and Oligocene deposits, although found in all three major oceans, are scarce. Eocene, on the other hand, is widespread and important. A new and abundant phase of clastic sedimentation appeared in the Miocene, probably following widespread orogenesis about the end of the Oligocene. Pliocene, being superficial, is the most common pre-Quaternary sediment — consonant with the widespread denudation of the lands at this period (p. 129).

The distribution of individual stages is only now becoming known from acoustic reflection studies, and certain strata are shown to be more uniform and extensive than might have been anticipated, the strength of bottom currents being stronger than was specified by authorities a generation or so ago.

The new profiler and coring data sometimes effect great simplification of regions formerly thought to be tectonically complicated. The Caribbean data, for instance, are summed up by J. Ewing *et al.* (1967) as follows:

'Seismic profiler surveys and sediment coring have indicated that the interior basins of the Caribbean Sea have been stable, deep water areas since middle Mesozoic and possibly earlier. The mapping of a widespread marker of late Mesozoic or earlier Cenozoic age permits certain inferences about the tectonics of the margins of the basins. Vertical, rather than horizontal, forces appear to have been dominant in the deformation. The small variation in the thickness of the Tertiary sediments argues against the concept that convection, producing sea-floor spreading, within the Caribbean Sea is responsible for the marginal deformation.'

A section across the Beata Ridge illustrates the verticality of local displacements (Fig. 32).

Sediments on the mid-ocean ridges and in adjacent basins

Sediments are almost absent from the crests of ridges and basaltic crust is frequently exposed there. This, with the presence of a rift valley at the crest, is a plain indication that the crestal zone at least has undergone late Cenozoic to Recent formation with volcanism and some erosion (for occasional patches of sediment, mainly Miocene, are found almost up to the crestal rift at some places).

On the ridge flanks within 400 km from the axis the thickness of sediment gradually increases, and this condition indicates the presence of an earlier, broader cymatogenic stage — perhaps mid-Cenozoic. Beyond the 400 km point the flanking sediments are suspensoids in strata of almost constant thickness, or they thicken only slightly with increasing distance from the ridge crest. This indicates that the ocean basins had been formed, and continental drift virtually completed, before a long period of quiescence (probably early Cenozoic) during which most of the observed sediments were deposited uniformly over the main oceanic basins, e.g. Atlantic. Here samples collected showed, up to 800 km from the axis, either Pliocene or Miocene; beyond 800 km only one Miocene sample and one Cretaceous, the rest were either Eocene or Pliocene. Oligocene was absent.

Rona (1970) concluded: 'The comparisons suggest that, to a first approximation, the mean subsidence rates of the opposing continental margins and the mean spreading rates of the intervening sea floor have varied in unison during much of the Mesozoic and Cenozoic. This apparent covariation is consistent with the stratigraphic and seismologic data cited which indicate that the opposing continental margins and adjacent basins have behaved as if vertically coupled.'

'If continental crust is coupled horizontally to oceanic crust in the area investigated, it is obvious that convective transfer of material could not involve differential movement between the two. Any convection beneath the continental margin would be subcrustal.' And later: 'On both sides of the Atlantic coastal plain and continental slope, seaward thickening wedges of Cretaceous and Cenozoic strata dip seaward regionally at less than 5°'. Their wedging and greater dip in older formation, demonstrate the presence of coastal monoclines and intermittent deepening of the ocean basins (see p. 140).

A number of acoustically reflective horizons are present amid these basin sediments, and the extent and uniformity of these horizons confirm both the relative quiescence of the sedimentation, and the prior establishment of the oceanic basins. They are frequently valuable for long-distance stratigraphic correlation (p. 140).

The sedimentary history of the ocean basins tells of: (a) rapid dispersal of the lands by continental drift during the Mesozoic; followed by (b) long-continued quiescence during the early Cenozoic; with (c) renewed tectonic activity, largely vertical in expression, during late Cenozoic and Quaternary time. Thus Ewing and Ewing (1967):

'1. An earlier cycle (or cycles) of spreading occurred, possibly that during which Gondwanaland and Laurasia were broken up and the pieces moved approximately to their present locations. The spreading during this cycle was extensive enough to sweep all of the Palaeozoic sediments from the Pacific floor. Its duration was probably considerably shorter than the following period of quiescence. Reversals of the earth's magnetic field were recorded as bands of anomalies symmetrical about the ridge axes, as postulated by Vine and Matthews.
2. The early spreading cycle terminated probably in the late Mesozoic or early Cenozoic and was followed by a period of quiescence, during which most of the observed sediments were deposited on a static crust.
3. A new spreading cycle commenced about 10 million years ago, which has generated the pattern of the magnetic anomalies of the crest and has produced the strip of thin sediments. The ridge axes of the latest cycle follow those of the preceding one with remarkable accuracy.'

The sea-floor spreading terminology apart (which may now be out of date), the history is true also of topographic development upon the lands (Chapter 10), where it is known in much greater detail (King, 1962, 1976), and it should be compared with the independently derived table of sedimentation on p. 129.

A fundamental relationship is preserved in the attitude of sediments relative to the ridge. Two model cases are: (a) if and where the sediments are warped up towards the ridge, without loss of stratal thickness, postsedimentational vertical elevation of the ridge area is indicated. Alternatively, (b) if the strata overlap the flanks of the ridge and thin out successively towards the axis of the ridge, (Fig. 33 and cf. Fig. 29), then the relationship is possibly that of normal filling of a basin with each successive horizon overlapping on to the ridge flank. Then the ridge and basin will both be antecedent to the sedimentation, and drillings will show successively older rocks resting upon basement as distance from the ridge axis is increased. This would be by normal basin filling and overlap. *No spreading of the sea floor is expressed in such a relationship* — although it is often quoted as the prime demonstration of sea-floor spreading! All it may demonstrate is that the deepest parts of the ocean basins (not including the recent linear or arcuate ocean deeps) were the first to accumulate bottom sediments.

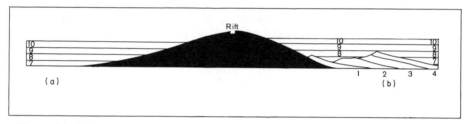

Figure 33   Diagrammatic section showing two types of stratigraphic relationship found on the flanks of suboceanic ridges. (a) Tectonically quiet overlap and basin filling. The age of the sedimentary formation overlying the risen ridge is progressively older the farther the formation is away from the crestal rift. Such a relationship has sometimes been claimed as evidence for sea-floor spreading. (b) Unconformity and older strata uparched towards the ridge crest. The basal layer is uniformly of the same age and it does not thin out by overlap. The strata below the unconformity are older than the uparching of the ridge

Similar stratigraphic relations appear in sediment-receiving basins upon the lands (e.g. Kalahari) where the most youthful sedimentary infill appears highest upon the basement slope.

In other words, if continental drift is admitted, it would appear to have been a relatively rapid event followed by long-continued quiet sedimentation within the ocean basins.

Both models of sedimentation have been encountered repeatedly upon the ocean ridge system, and sometimes both models are present in the same locality. These occurrences demonstrate intermittent rise of the ridge during sedimentation. The vertical uplifts are most commonly dated at late Oligocene–early Miocene and pre-Pliocene. Quaternary movements are not excluded.

Large rivers sometimes build great cones of land-derived detritus far into the ocean basins. These act isostatically as considerable loads upon the rigidity of the oceanic crust. The load of the Amazon cone (Cochran, 1973) 'may be taken to indicate a crustal thickness of only 30 km which is small compared with the 100 km indicated by seismological considerations.' In the Indian Ocean the cones of the Indus and Ganges Rivers make equally large submarine features. The combined volumes of these two cones comprise almost half the entire volume of sediments in the Indian Ocean. The basins adjacent to Africa amount to about half of the remainder, with a small addition from Australia.

In the Pacific Ocean refraction profiles and sediment cores show the sediment accumulation to be thinner than in other oceans. For this there are several reasons:

(a) The Pacific is ringed by mountain chains and island arcs so that the area shedding clastic debris to this ocean is small by comparison.
(b) Much of the Pacific is bordered by ocean deeps which act as sediment traps.
(c) The area of ocean is greater than all the other oceans combined, and its floor over which the sediment is spread is correspondingly greater. Most parts are remote from the continents and the sea-floor deposits 'show remarkable simplicity and continuity in the sediment layering through all parts of the North Pacific except those adjacent to island platforms or ridges.' (Ewing *et al.*, 1968)

The only river which has built a large sea-floor cone into the Pacific is the Columbia River of Oregon.

The main bulk of sediments in the Pacific is therefore biogenic and pelagic in origin and their thickness distribution follows that described earlier in this chapter. This pattern of sediment distribution was the same throughout the Cenozoic as at the present day; but the Mesozoic accumulations are apparently most abundant in the northwest Pacific basin, e.g. Shatsky Rise.

Northeast Pacific profiles show thicker sediment in that sector presumably because of abundant turbidites from North America in addition to the normal lutite deposition. Turbidites south of the Aleutian Trench correspond with turbidites from the north. The sinking of the trench must therefore be a geologically recent event. More than one model has been presented under the hypothesis of sea-floor spreading and plate tectonics, and these differ greatly in estimation of the time for sinking of the trench from 25 million years or more down to as little as 7 million years ago.

Turbidites are common also in the northwest Pacific, decreasing in thickness eastward. As the Kurile island arc would have been insufficient here to supply the alleged turbidites, either the identification of the sediments as turbidites must be wrong, or again the trench is very youthful. For comparison, much of the sediment in the Shikoku basin east of Japan is derived from the Japanese mainland although the area is one of high biological productivity. These sediments are well stratified, nearly horizontally bedded and are thought to contain many turbidites.

In mid-Pacific the Darwin Rise (as defined by Menard, 1964) includes Polynesia, Micronesia, and sundry adjoining areas. Not only is the area studded with chains and groups of islands but it 'contains a large percentage of all the atolls and guyots in the ocean basins'. All these features are volcanic and were mostly in eruption between 60 and 100 million years ago, These seem to have supplied quantities of Cretaceous volcanogenic sediments which thin locally away from the volcanic sources. Manihiki Plateau (162°W, 11°S) is typical, and there (Jenkyns, 1976) DSDP boreholes 317, 317A and 317B show it to have been formed of 'abundant outpourings of basalt that grew into shallow water'. Reefs formed at the same time (early Cretaceous) and the whole assemblage was rapidly acted upon by sea water to produce characteristic clays. Then there was rapid sinking. Turonian and Santonian sediment are absent and there was apparently erosion. The Cenozoic to Recent record was dominantly calcareous.

The youngest chert present is Oligocene. Chert Horizon B (p. 134) has also been identified in these sediments. Cenozoic pelagic sediments are much thinner than usual owing to very slow rates of accumulation. No other source for the sediments seems reasonable. Other terrigenous areas are too far away to have supplied turbidites, and the Darwin Rise itself appears to have been a shallow area at the time of its volcanic activity. The Darwin Rise region does not seem to afford evidence of sea-floor spreading during Cenozoic time; but it supplies large-scale evidence of vertical displacements ranging from near sea level to a depth of 2000 metres at the present day.

Leg 17 of DSDP also affords precise information on this Central Pacific area. Site 164 at 13°12' N, 161°31' W where the water depth was 5513 metres, was drilled through 278 metres of sediment into 10 metres of vesicular basalt. The uppermost 50 metres of sediment, very soft, was accoustically very transparent and so soft that no samples were taken. It 'graded down into Oligocene radiolarian, brown zeolite clay. The first cherts appear at 50 metres in the lower Oligocene and continue to the base of the sedimentary section in the lower Cretaceous. Zeolitic brown clay with sparse Radiolaria is interbedded with the chert. Aside from a few coccoliths (lower Cretaceous) no calcium carbonate was detected.'

Site 165 at 8°10' N, 164°51' W was drilled for 490 metres beneath a water depth of 5053 metres. Sited 50 km from a guyot in the Line Islands the upper 240 metres of core was 'Eocene to lower Miocene calcareous turbidites interbedded with radiolarian ooze, then 50 metres of interbedded Eocene and upper Cretaceous chert and limestone, 140 metres of upper Cretaceous volcanogenic turbidites and 50 metres of alternating basalt flows and volcanogenic sediments resting on basalt. These results suggest that the Line Island seamount chain is somewhat younger than its surroundings . . . and that portions of the chain have been in shallow water during Eocene and late Cretaceous times.'

Conclusions by the scientists of Leg 17 include:

(a) 'The pattern of the ages of the basal sediments does not fit a simple extrapolation of age gradient determined from magnetic anomalies and drilling results east of the Line Islands. This probably indicates that the crust of the Central Basin was generated from a different spreading axis, as is sugge-sted by the east–west anomaly pattern recently mapped. . . . Alternatively, it must be assumed that a major volcanic period during middle–late Creta-ceous time buried the oldest sediments at several of the sites. In most of these holes flows, sills or ash layers gave evidence of major volcanic activity during this period.'

(b) 'The very slow rate of accumulation of early Tertiary sediments at these sites, and at others in the Atlantic and Pacific, suggests a major environ-mental change of global extent. . . . At some sites, e.g., 166, the hiatus extends well down into the Cretaceous. Hard middle Eocene cherts are typically encountered in the section at, or just above, the zone of slow accumulation.'

(c) 'The drilling on Horizon Guyot gave conclusive evidence that parts of the mountain stood above sea level during middle Cretaceous time.'

(d) 'The deepest sediments at sites 169 and 170 are calcium carbonates, which may indicate a substantial subsidence of the basin, or a deepening of the compensation level.'

(e) 'Where there was no ambiguity due to incomplete sampling or uncer-tainty in dating, seismic Horizon A (base of the upper transparent layer) correlated well with the top of a zone of hard cherts of middle Eocene age that lie just above the Early Tertiary hiatus.'

These conclusions are among the most important yet won from the ocean depths remote from land. Their chronological significance, and possible correlations with events upon the land masses, appear in the following chapters.

In the South Pacific are areas of specially thin cover over which are high concentrations of manganese nodules. These characteristics indicate (Menard, 1964) a very slow deposition. One is tempted to equate these occurrences with the long period of quiescence during the early Cenozoic observed both in the oceans and upon the lands, where laterites developed upon the Moorland planation (p. 188).

In the southwest Pacific deep-sea sediment shows several changes of regime during the early Cenozoic, as though induced by tectonic changes of sea-floor bathymetry. This is especially so south of Australia and New Zealand; but following regional unconformity in the late Oligocene, modern distribu-tion of sediment was established. Once more we note early Cenozoic siliceous biogenetic strata upon the western edge of the Campbell Plateau and south of Tasmania. These possibly signify deep cold water, and are succeeded by widespread calcareous facies in the Neogene (when glaciation became widespread in Antarctica). Perhaps beginning in the Miocene there was much movement to north and south of the Antarctic convergence (e.g., the convergence presently lies (abnormally) north of South Georgia which, at 55°S, is truly Antarctic).

Observation of the continuity and extent of sediment over the Pacific (and

136

other ocean basins too) does not seem generally to favour the hypothesis of sea-floor spreading. Indeed it often renders the hypothesis superfluous. The same pattern of biogenic deposition, for the Pacific particularly, has persisted from the beginning of the Cenozoic to the present day. But diachronism within the zonal deposits has been explained by Heezen and Macgregor (1973) in terms of a west northwestwards travel of the sea floor across the equator (Fig. 34).

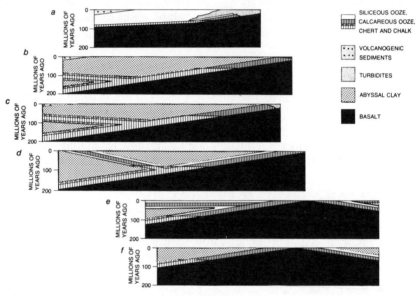

Figure 34   East–west cross-sections of sediments beneath the northwest Pacific Ocean illustrating the diachronous relationship claimed as due to northwest travel under plate tectonics of the ocean floor across the equator (after Heezen and MacGregor, 1973)

Interpretation of sedimentation in the evolution of the northwest Pacific basin

The northwest Pacific affords perhaps the most satisfactory area of enquiry for the evolution of oceanic basins because:

(a) It is the largest of existing oceanic basins.

(b) In the west it has the oldest existing formations (Jurassic) of the post-Palaeozoic marine sediments.

(c) It has no median submarine ridge, but has a modern cymatogenic ridge on the eastern side only, giving the greatest area of ocean floor known between sea-floor ridge and boundary continent (Asia).

(d) In the west it has some of the deepest troughs of the ocean floor.

(e) It has the minimal terrigenous sediment and the greatest volume of pelagic sediments anywhere in the oceans. This reduces the influence of continental tectonics and denudation upon the sedimentary record.

(f) As the eastern side includes the East Pacific Ridge, a zone of crustal upwarp, and the western side is festooned with marginal seas, island arcs and ocean deeps signifying a region of progressive Cenozoic sinking, the whole makes a crustal plate with progressive tilting from east to west.

(g) It is not a simple extensional basin postdating the break-up of Gondwana and Laurasia, but was an oceanic area before that cataclysmic event.

(h) With the above variety of features, the region may furnish information not available in other oceans, and provides therefore a better example of standard ocean basin evolution: witness, for example, the Geological Map of the Pacific Ocean by Heezen and Fornari (1976) whereon west of longitude 150°W the map looks like a typical geological map, but east of this line (approximately) it looks like a geophysicist's model — which is not the same thing. The timing of the difference is upper Cretaceous.

'The sea-floor of the northwestern Pacific is covered by five stratigraphic units: (a) a late Tertiary silty clay, primarily of volcanic origin, existing as a wedge thinning away from the volcanic arc source; (b) a late Cretaceous to early Tertiary zeolitic clay, which is the normal abyssal clay covering the whole area as a layer 10–50 metres thick; (c) a sequence (about 100 metres) of Cretaceous chalks and cherts; (d) an inferred earlier mid-Cretaceous thin abyssal clay; and (e) late Jurassic to Cretaceous chalks, cherts, limestones and marls (more than 200 metres). The underlying basement is Late Jurassic or older.'

The northwest Pacific region was the subject of investigation on Legs 9 and 16 of DSDP which provided a wealth of new data which were interpreted by the original investigators at sea in terms of the sea-floor spreading hypothesis* from whom we quote. The report on sediments on the west flank, East Pacific Ridge, is as follows (Figs 34, 35):

'Three principal factors govern the depositional history of the eastern equatorial Pacific —

(a) Sea-floor spreading moves the crust westward from the shallow crest, where carbonates accumulate, into the deeper water of the lower flank, which lies below the calcite compensation depth and received only biogenous silica and some inorganic detritus.

(b) A general northwestward shift of the Pacific plate, which appears to have continued since the early Cenozoic, carried the sea-floor from a zone of slow deposition in the South Pacific through the equatorial zone of high productivity and rapid deposition into the north-central Pacific where deposition is again slow. The resulting sequence of deposits changes from siliceous to calcareous to siliceous.

(c) Finally, the calcite compensation level — the depth below which calcite is completely dissolved — moved from a relatively shallow position during

* *Geotimes* (June 1971, **16,** pp. 12–14).

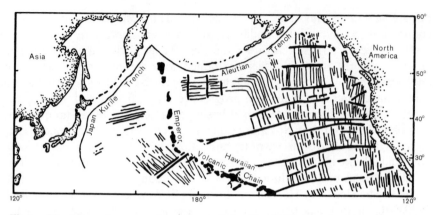

Figure 35   Magnetic reversal patterns of the North Pacific plate in relation to the coast of North America, the East Pacific megashears, the Aleutian Trench, the Kurile Islands, Japan, and the Emperor-Hawaiian chain of volcanic islands and seamounts. Several discordances are apparent and tectonic episodes of several different ages are probable (after Hayes and Pitman, 1972)

the Eocene to a depth in the Oligocene at least 1000 metres greater than the present 4600 metres or so. The interaction of these three factors has determined the arrangement in time and space of calcareous (rapid deposition above compensation depth) and red and brown clay (slow deposition below compensation depth) facies in the equatorial zone.'

Thus all three of these factors may be explained without lateral crustal shift at all, merely by vertical up and down displacements, local and regional, of the type common upon the lands and without doubt equally so beneath the ocean. Menard (1964) has described them for the central Pacific as a huge bulge of the subcrustal mantle (the Darwin Rise) during Cretaceous time. There the mid-Cretaceous Sea was replete with banks, islands, and active volcanoes already showing sands and cobbles as the upstanding masses were worn away. Corals and other calcareous shelly creatures grew abundantly to furnish carbonate deposits above the compensation depth. Then, early in the Cenozoic, the whole vast structure began to subside, which it has continued to do to the present day. So between the Jurassic and the mid-Cenozoic would be developed the triple sequence — siliceous –calcareous–siliceous — found in the core records. And as the Pacific basin has tilted from east (where the cymatogenic ridge has formed a zone of uplift) to west which is marked by submergences and a general greater depth, there would be a natural accretion of carbonate deposits in the shallower eastern waters.

'D.S.D.P. Sites 159–162 also show the effects of westward sea-floor spreading, both by the increasing age of deposits immediately overlying the basement (which is extrusive in all cases), and by a westward reduction in carbonate content in each stratigraphic interval as the sea-floor grew progressively deeper.' These effects are equally explicable as the result of eastward stratigraphic overlap following progressive westward tilting of the sea floor, both bathymetrically and with respect to a timeous compensation level. The sedimentary records should certainly not be taken to demonstrate 'spreading' of the Pacific basin floor, to the exclusion of other simpler hypotheses — for which is simpler(?): to translate such huge areas, volumes, and masses of suboceanic crust as are involved here over distances of thousands of kilometres and for scores of millions of years (as sea-floor spreading prescribes), or to visualize the thin crustal blanket as heaving and tilting both regionally and locally above an uneasy fluid mantle seeking direct outlet for its global energies (p. 122).

A cardinal fact adduced by sea-floor spreaders is that the age of basal sediments resting on that crust becomes progressively older the farther they are away from the ridge axis; from which they *assume* that the sediments (and the upper crust upon which they rest) have necessarily been moved laterally away from the axis of a mid-ocean ridge.

But the same age relationship is achieved by normal overlap on to the flanks of the mid-ocean ridge by sedimentary filling of the adjacent basins (Fig. 33), and there is also evidence that the basins have themselves subsided periodically during their existence (p. 207). Only *vertical* movement is necessarily involved — cymatogenic elevation of the ridge coupled with related subsidence of adjacent basins. The sedimentological evidence alone does not demonstrate clearly either horizontal movements of the oceanic crust, or upper mantle thermal convection cells.

Perhaps a solution lies in a combination of horizontal movements with elevation of the type found in the Darwin Rise?

Chronological importance of sediments

Sediments, often of marine origin, upon the lands have yielded a chronology of the utmost service to geologists for the past two hundred years. With deep-sea drilling the opportunity arises to study marine deposits in their own environment, to ascertain the proportions of pelagic and terrigenous materials, and to learn what information they afford on the origin and development of the ocean basins. How are they related, for instance, to: (a) the tectonic mid-ocean ridges; and (b) to the continental margins (p. 207). How extensive are individual formations within the ocean basins? Information both regionally and sequentially is invaluable.

Samples of marine sediments carry their own datings through the fossils that they contain. Of recent years they have been found also to exhibit geomagnetic polarity reversals representative of their stratification. These reversal datings should be read stratigraphically (vertically) rather than laterally, so long as the present queries on the validity of sea-floor spreading remain.

The extent of individual formations revealed by seismic reflection devices is sometimes surprising — they may cover many thousands of square kilometres with relatively little tectonic disturbance. And, of course, where they occur they indicate that ocean basins existed at the time of their accumulation — even if the sea-floor spreading hypothesis and the geomagnetic polarity reversals suggest that a particular basin area did not develop until later!

Formations of very uniform, fine-grained composition (e.g. cherts) are typical of this class. They represent tectonic quiescence with low denudational rates upon the lands during their long, slow accumulation in the oceans.

Horizons 'A' and 'B'

Extensively beneath the western Atlantic basin, under 300 to 500 metres of younger sediment, is a strongly reflective layer (Horizon A) which marks the top of a turbidite sequence making an abyssal plain of late Cretaceous age covered by an early to mid-Eocene chert.* *Horizon A extends from beneath the continental rise in some places to the base of the mid-Atlantic Ridge.* The horizon does not now extend as a blanket over the ridge. In 1977 Dr R. E. Houtz wrote me from Lamont that Horizon A consists of four different formations with the cherty reflector being the only one that is widespread in the Atlantic.

In the west the same chert is let down in tectonic blocks into the Puerto Rico Trench, and beyond that a similar reflector is widespread in the Caribbean basins, where it was examined on DSDP leg 15. In the middle of the Venezuelan Basin sites 146 and 149, beneath 3949 and 3472 m of water, Horizon A was obtained between 406 and 440 m of a 746 m core section ending in 16 m of dolerite. Fauna and flora are almost entirely planktonic and volcanic ash is almost the only detrital material. The age may be late Paleocene. Upper Cretaceous marls and chalks below appear allocthonous as by turbidity currents. Above Horizon A (149) is a relatively complete succession of Tertiary beds topped by 40 m of Pleistocene.

Site 147, sunk in the Carioca Trench on the continental shelf of Venezuela, did not reach Horizon A.

Site 151 on the southern part of the Beata Ridge found that Horizons A and B approach more closely than usual, an indication that the ridge had risen into existence between the mid-Cretaceous and the mid-Eocene. Normally Horizon A shows no folding over vast areas of the ocean floor; so it demonstrates clearly that some 2000 km width of the western Atlantic basin was in existence shortly after the end of the Mesozoic, and by the early Cenozoic already was composed of immense depositional abyssal plains. This leaves little space in the west Atlantic for any later Cenozoic sea-floor convection belts.

* Chert is considered to be the lithified product of abundant marine organisms with skeletons of opaline silica. Often the concentration of dissolved silica in interstitial water of marine sediment is considerably higher than in the supernatent water at the ocean bottom. This particular example was dubbed *Horizon A* by Ewing.

On the eastern side of the ridge many strong reflectors are present in the North Atlantic, rendering individual recognition of Horizon A sometimes difficult. But the extent of sea floor in which these reflectors appear is again such as to indicate that most of the eastern half of the Atlantic also was in existence early in the Cenozoic, and the space available for Cenozoic sea-floor spreading is embarrassingly small.

In the South Atlantic basins, too, a similar reflection to Horizon A is widely identifiable. In the Argentine basin it is overlain by thick layers of homogeneous sediments extending from the continental slope to the flanks of the mid-ocean ridge. These are of terrigenous origin and were transported from the continental shelf out into the ocean basin as vast clouds of murky water carrying quantities of fine-grained lutite in suspension. Reflecting surfaces within the sediment are practically level and undisturbed throughout the length and breadth of the Argentine basin, which was evidently formed during the Cretaceous Period. The reflectors, if truly identified, allow little space for sea-floor spreading during the Cenozoic or Quaternary.

In the eastern South Atlantic, the Orange River Delta which began to form in the Cretaceous Period includes a reflector thought to be Horizon A; and there is a further identification of Horizon A at 57°S, 30°E amid turbidite beds like those of the Argentine basin. So the South Atlantic basins appear to have extended into the area of the Great Southern Ocean.

In a few tectonically disturbed areas Horizon A attains a relief of up to 1.5 km. This is exceptional; but on one such elevated area near the Bahamas, sea-floor erosion has exposed a small area of Horizon A from which the first samples showed it to be an Eocene chert underlain by Cretaceous layers. Since then improved techniques of deep-sea drilling in alternately hard and soft rocks have frequently penetrated it.

Other ocean basins possess an analogous strongly-acoustic reflector. Over a large part of the southwest Arabian Sea in DSDP boreholes 219, 220, and 221, there was determined a lower to middle Eocene chert which shows: (a) that a large part of this region was under the sea early in Cenozoic time; and (b) that any later sea-floor spreading must have been minimal. On the other hand the Arabian abyssal plain occurs generally beneath 4000 metres of water, proving that later vertical subsidence has been considerable, and the corresponding history of sedimentation is almost continuously pelagic ooze from late Miocene to Holocene.

The 'essentially non-magnetic' Owen Ridge generally shows an asymmetric cross-profile with a gentle western slope dipping beneath an abyssal plain, and a steep eastern fault-scarp holding back Indus Cone deposits. After an hiatus, renewed uplift in the late Miocene was inferred by investigators of benthonic foraminiferal assemblages found in the cores.

The dates mentioned, and the vast cones of sediment deposited by the Indus and Ganges Rivers, correspond with the mid-Cenozoic rise of the Himalayas and the development of the large rivers which later cut antecedent gorges through the rising mountain chains.

More and more acoustic reflector profiles (echograms) are being identified, and the continuity with which individual recognizable horizons are present in the sea-floor sediments shows that the ocean basins have not grown slowly through geologic time, but that there were Mesozoic phases of great activity with rapid ocean widening, followed by long intervals of quiet, widespread deposition of turbid or pelagic materials through the early Cenozoic. There is a marked rhythm of global tectonics detectable in the ocean basins, as upon the lands (p. 202).

Even the Pacific, which is an ancient ocean, is not exempt. In the basins of the northwest Pacific Horizon A is almost ubiquitous (Ewing *et al.*, 1968). After leg 6 DSDP drillings it was recognized that the upper surface of the layer varied in lithology and somewhat in age. Just west of Hawaii the reflector was identified in middle Eocene chert associated with chalk in one area and with mid-Tertiary volcanic ash in another area. Even in the Central Pacific between Hawaii and Tahiti, where most of the *Glomar Challenger* 'cored material was calcareous nanno plankton ooze', core terminated in Eocene indurated sediments and chert, or in basalt. These are open-ocean equatorial sediments, not like the turbidites of the Atlantic; but the typical Eocene chert is present. Perhaps it is due to a sojourn of the sea floor below the level of 'carbonate compensation' during the early Cenozoic; but the Eocene age once more is significant. In the northeast Pacific floor, also, a layer was tentatively identified by Ewing *et al.* (1968) as Horizon A (Eocene).

Although Eocene ages are frequently cited for these chert beds in several oceanic basins they need not be regarded as precisely synchronous. The top of the chert layer is widely time-transgressive, and the deposit is not infrequently multiple with soft intervening strata.

When all the known areas of Horizon A are plotted it becomes evident that: (a) almost the full width of the Atlantic basins (both North and South) was in existence and filling up with sediment before the end of the Creta-ceous; (b) the hypothesis of continuous uniform sea-floor spreading from Jurassic to Recent is not tenable; (c) doubt is cast upon the hypothesis of geomagnetic polarity reversals in which the pattern of reversal stripes is interpreted as fixed within rock masses, instead of being free to pass through the static rock formations of the crust like a stress field; and (d) a long period of tectonic quiescence is indicated in all environments about the Earth during the early Cenozoic.

This correlates broadly (Chapter 10) with the great (Moorland) planations which developed over all the continents (King, 1962/7, 1976) during this same early Cenozoic timespan. Both in the oceans and upon the lands examples differ somewhat chronologically according to local influences; but though the earliest may be Paleocene and the latest Oligocene, there is marked similarity and dates congregate in the lower to middle Eocene bracket.

Furthermore, (e) in marine environments there is a tendency for chert reflectors to form at those times when upon the land masses the planed landscapes develop laterites, bauxites, and their congeners (Chapter 10).

Sediments below Horizon A are of variable thickness and rest sometimes upon a flat terrain; sometimes upon a rugged one. They are thought to include a Cenomanian acoustic reflector, Horizon B (which is again chert); and they extend in many places down to, but seldom beyond the early Cretaceous. Horizon B is found in both the Atlantic and Pacific basins where it is sometimes in contact with the basement. This horizon may be distinguished from basement 'primarily by its characteristically smooth upper surface'. The surface of the basement is normally rough and dredged samples show it to be basalt or ultrabasic rock.

Ewing dated Horizon B as Cenomanian. This dating is said to be in some doubt; but it fits very well with the geomorphology of the continents where it agrees with the Kretacic planation (King, 1976) developed during the early and mid-Cretaceous (Chapter 10).

In a letter dated December 1977, Houtz reported: 'Horizon B was the name used for "smooth acoustic basement" as it appeared in reflection records when a small airgun was used as a sound source. The use of much larger airguns reveals in many places that it is actually a seismically reverberant layer composed of calcareous, volcanogenic, or cherty materials that were "opaque" to the small guns. Accordingly, the distribution of Horizon B is not a consistent geological event.'

Chapter 9

Chronology of the ocean basins

Postwar research in the ocean basins

The forms and structures of the major oceanic basins have been made clear only since World War II, first by the programmed researches of the Lamont Geophysical Institute under Dr Maurice Ewing, and lately by the extended programmes of deep-sea drilling by research vessels (e.g. *Glomar Challenger*). The contributions to knowledge made by these (and many other) oceanographic institutes outweigh those of the exploratory moon missions in variety of techniques applied and in the novelty and value of the results achieved.

Because the techniques of remote sensing employed have been mainly physical (with physical interpretation of the results) there has been some tendency to neglect: (a) geological assessment of the same data; and (b) to integrate it with geological information from the lands to achieve a global synthesis of geology. The first paragraph of this book emphasized the relationship between supramarine and submarine topographies; we shall therefore: (a) review certain of the new data available; (b) assess new techniques of research from the standpoint of a geologist; and (c) in later chapters compare the new topographic and related data with known landforms and endeavour to relate them globally as earthscapes. (d) We do not attempt a compendium of oceanographic data.

The chief advances made by this spate of oceanic research are:

(a) Through the echo-sounder. The new data have revealed a more detailed bathymetry of the ocean bed so that, in some offshore areas (e.g. around the United States) the submarine topography has been defined with almost visual accuracy.

(b) Improved methods of sonic sounding now yield records which provide continuous sections through the bottom sediments, which makes their succession and structures clear over long traverses (Fig. 49). The thickness of specific formations may be faithfully rendered, and the top of the basaltic crust usually concludes the sequence, a useful datum indeed.

(c) Seismic refraction and reflection patterns afford further information of the suboceanic lithosphere, revealing its composition, density, succession, and structure. Intrusive structures are often clear.

(d) Heat measurements are more easily made upon the ocean floor than on the lands, and hence are more numerous at sea. Enhanced heat flows are revealed at plate boundaries and occasionally elsewhere in relation to the rise of volcanic materials.

(e) Patterns of magnetic polarity reversals (p. 97) are revealed.

We shall now consider mainly those data which are germane to the topics discussed in previous chapters.

Distribution, forms, and age relations of the major suboceanic ridges and basins of the Atlantic and Indian Oceans

Modern opinion on the relation of continents to the great suboceanic ridges derives from the North Atlantic, where the North American and Eurasian continents are thought to have withdrawn progressively from the site of the mid-Atlantic Ridge since early Cretaceous time. Morphologically, the Atlantic continental margin of North America is typical, with a broad, flat continental shelf, a smooth continental slope and an even flatter continental rise. Faulted marginal basins contain 8–12 km of late Mesozoic to Recent sediments. 'Mantle depths are consistently 12–15 km under the outer rise, about 20 km under the slope and 30 km beneath the shelf.' But these figures do not narrowly define continent–oceanic crustal boundaries. In the coastal plain and hinterland marine formations are exposed which permit a correlation of the major continental events with the offshore sedimentary record (Chapter 11). These events are generally synchronous (late Cretaceous, early Miocene, Pliocene, and Pleistocene) with those of other monoclinal coasts (pp. 205–207).

Another general observation is that the basins of all the oceans (and to some extent continental basins also) display a history of progressive and/or intermittent subsidence. This is so for the North Atlantic, with much faulting and deepening around 50 million years ago. The only surviving plateau areas are the Newfoundland Banks (submerged) and Rockall, which has almost entirely disappeared; nowadays its granite cliffs are virtually inaccessible amid the rough seas.

The basins of the Atlantic contain more than half of the world's abyssal plains, for sedimentation is maximal in this ocean. The abyssal plains are formed by the ponding of sediment over long periods of time, e.g. Argentine basin.

The northern hemisphere has most of the land areas of the globe, and most of the people. Stratigraphical study began there (c. 1800 AD) and it was not until 150 years later that the vital contributions of southern lands to global geological thought were understood, with acceptance of the Gondwanaland concept and continental drift.

A similar position has arisen with the exploration of the oceanic basins. Geological and geophysical studies began in the North Atlantic, and the interpretation of sea-floor evolution in general has followed upon that simple model. Now the more numerous and complex series of basins and ridges in the southern hemisphere pose new problems and invite new thinking.

Despite conclusions drawn from the North Atlantic, Chapter 3 found that the major submarine ridges are of vertical cymatogenic nature; and Chapter 5 has shown that neither the world's submarine ridges nor their positions in the oceans were caused by tension between oppositely directed conveyor-belts in the asthenosphere, which is assumption number one to students of the North Atlantic.

The southwest Indian Ocean Ridge, for instance, has no crestal rift and not much volcanism, but seismically it has as much activity as a symmetrical ridge of the mid-Atlantic cymatogenic type. Green (1972) has indicated that it does not even exhibit a regular succession of geomagnetic polarity reversals; but this may be due to two mighty fracture zones (Agulhas and Prince Edward) trending north-northeast with strong dextral shear. East of Mada-gascar the Mauritius, Vishnu, and Mahanoro fracture zones trend northeast to southwest with similar rotation. These, and other, fractures are dominant in the sea-floor structure here (Norton and Sclater, 1979).

Another example: the largest ridge of all is the Antarctic–East Pacific Ridge. But why does the Pacific have a ridge at all? The basin of the Pacific is not a disjunctive basin as the Atlantic is claimed to be, with progressive withdrawal of its bordering continents to either side away from a median ridge. The borders of the Pacific have everywhere encroached *into* that ancient ocean basin. Those confining coasts are everywhere *inherited* portions of the ancient coasts of Laurasia and Gondwana evolved in, at least, late Proterozoic time. Certainly some of the ramparts have been rebuilt by Ceno-zoic mountain-making; but there is nothing like the fractured coasts, fringed with coastal plains, which border the Atlantic.

In the 9000 miles from the Antarctic to Easter Island the East Pacific Ridge has no axial rift though its shallow earthquake foci are normal for a ridge crest. However, to the east it is separated from South America by a zone of ocean deeps and the evidence of a subductive structure is completed by the abundant records of deep-focus earthquake foci beneath the Andes. (A situation which is mirrored in the West Pacific by the Tonga–Kermadec trough with associated volcanism and seismicity.)

But though the ridge here lacks a rift, the geologic map of the Pacific by Heezen and Fornani indicates clearly that the longitudinal shear pattern has numerous cross-shears (some expressed vertically) that we found (p. 169) typical of the stress measurements of Hast (Fig. 42).

Circum-Pacific orogenesis cannot be caused purely by Cenozoic sea-floor spreading, as some authors have claimed, for the earliest orogenesis known in those coastal regions was Precambrian and was related to the margin of Gondwana.

Menard and others early specified differences in the form and functioning of the Pacific Ridge at intervals along its length which indicate that the ridge is a compound, agglomerative structure? The modern Antarctic–East Pacific Ridge (p. 169) has grown up and achieved united tectonic form only during the past 9 million years (Herron, 1972).

Each ocean has a submarine ridge prominent in its bathymetry. All are currently active, both seismically and volcanically, with enhanced heat flow. A crestal rift and transverse fractures are commonly present, and the tectonic activity now displayed seems to have begun in many regions about the beginning of either: (a) the Cretaceous period (e.g. mid-Atlantic); or (b) the Miocene (e.g. Galapagos). But the ridges are not universally alike in detail, nor are they necessarily of similar origins.

Clearly, the southern oceanic basins, with the North Pacific, afford more scope for interpretation and testing of hypotheses than the North Atlantic alone can do. Indeed the evident symmetry of the Atlantic, both North and South, has perhaps misled geologists in their theorizing. No such symmetry exists in the other oceans. Plate tectonics helps in the interpretation of diverse phenomena, but attempts to find a common mode of formation for all basins, ridges, and other phenomena seem at present not to have followed up the earlier promise of Heirtzler *et al.* (1968).

Valid evidence of the opening of oceanic basins comes from the presence of new series of marine formations on the fractured and monoclinal, aseismic continental margins. These are the 'internal' coasts (Atlantic, Indian, and Arctic Oceans) produced by the rupture of Gondwana and Laurasia. Other, seismic coasts, around the Pacific sometimes with fringing seas and island arcs, are the coasts that were parts of the circumferential orogenic coasts of Gondwana and Laurasia respectively.

Deep-sea drilling in the Lesser Antilles records:

(1) from mid-Cretaceous to mid-Eocene 220 metres of sediment accumulated, slowly at first so that basement peaks still penetrated Horizon A (p. 140). These sediments thicken regionally towards the west.
(2) mid- to late Eocene saw the initiation of the Puerto Rico Trench which deepened during the Oligocene.
(3) during the Miocene the outer (Barbados) ridge built up.
(4) about the end of the Pliocene the Puerto Rico Trench attained its present dimensions.

The Blosseville coast of central east Greenland began to sink shallowly in the upper Albian, but not until the Eocene (Sparnacian) was a 9 km thick sequence of basalts extruded upon the earlier sediments. The Arctic Ocean basin opened up mainly during the Cenozoic, but then did so with considerable complexity (Fig. 37).

148

Figure 36 Polar projection of the Arctic showing main items of geological interest in the bathymetry. The ice-filled basin of Greenland possibly indicates what the Eurasian and North American polar basins were like during their development. The summit plateau of the Antarctic is similarly dished to below modern sea level, though the icy plateau itself stands 3000 to 3400 m above the sea.

To be noted is the almost continuous girdle of continental masses around the Arctic Circle. There is no long-continued scattering of continental masses as in the southern hemisphere (Fig. 31). Only the wide distribution of lower Carboniferous coals from Pennsylvania to Europe and on to China indicates accumulation under a uniform subtropical climate at one time, succeeded by later desert climates (Bunter) as the coal-forming period passed.

Over the same time interval Gondwana progressed (northward) from a south polar environment (Dwyka) to a cool-temperate (Ecca) Permian regime, and ultimately to a warm (southern) desert environment (Botucatu and Cave Sandstone). This covered the time leading to the collision between its northern parts and the southern fringes of Laurasia which was nudged northwards, as shown by plate tectonics along the Laramide thrust (Fig. 37)

Figure 37 Development of the Arctic Ocean structures under plate tectonics,
with the driving *northwards* of North American land masses by the Laramide
Orogeny, possibly caused by nudging from parts of Gondwana which have
crossed the equator. Solid black shows positions of surrounding lands at 81 Ma
ago; stippled are some areas in position at 63 Ma ago; pole of rotation in
Northern Greenland (after Herron, Dewey, and Pitman, 1974)

Geological similarities between Alaska and easternmost Soviet Republic
indicate that these two continental areas have been connected since the
Palaeozoic and probably since the Precambrian.

The story of the opening up of the Arctic Ocean, according to plate
tectonics (Figs 36–37) has been told by Herron *et al.* (1974). The pole of
plate rotations was sited on the north coast of Greenland and two main
basins were formed, called Eurasian and Amerasian respectively, of which
the former is the continuation of the North Atlantic. Between them three
ocean ridges cross from northern Greenland to the shallow continental shelf
of northern Asia: (a) the mid-Atlantic Ridge is continued by the Gakkel
Ridge towards the mouth of the Lena River; the Lomonosov Ridge is
described as the seaward edge of the Laurasian land mass prior to opening
up of the basin (63 Ma ago); and (c) the Alpha–Mendeleyev Ridge, which
is a Cretaceous subduction zone related to the Laramide orogeny.

The South Atlantic, of course, opened up a gash across the face of Gondwanaland. Although on fossil evidence the initial opening was in the north, the separation lengthened southward, the continent of South America, swinging on a plate tectonics pole in the north, ultimately opened up a much wider gap from Africa in the south (Fig. 38). The circle of fracture on which the two continents separated is the Falkland Plateau fracture which is carried across the mid-ocean ridge and is continued by the Agulhas fracture zone up the southeast coast of Africa to the latitude of Durban. Some 15° north of this fracture are two great submarine barriers (the Rio Grande Plateau and the Walvis Ridge) spanning almost the full width of the South Atlantic. The Walvis Ridge, an aseismic, non-volcanic sliver of earth blocks, stands athwart the eastern half of the ocean from Africa to the mid-Atlantic Ridge. Looking quite out of context, it stands (in apparent isostatic equilibrium) higher than the main mid-ocean ridge. It is said to contain complex faults and diapiric structures; and there is consensus of opinion that the Walvis Ridge has acted as a barrier to deep oceanic circulation since the South Atlantic opened during the early–mid-Cretaceous. The junction of Walvis Ridge with the coast of Africa coincides with a remarkable contrast of the coastal morphology and gravity information, and this difference seems to have been marked ever since the early Cretaceous when the South Atlantic opened. Continuous sea-floor spreading through the Cenozoic finds difficulty in explaining both this feature and the Bromley (Rio Grande) Plateau of the western South Atlantic linking to South America on the opposite side of the ocean.

The fracture zones of the equatorial Atlantic have a geomagnetic anomaly with a gravity high indicating probable intrusion of large quantities of ultrabasic rock from depth, associated with mantle expansion.

The coastal outlines of the Indian Ocean and their relative ages compare with the Atlantic and can be defined under the theory of continental drift; but the geographical relationships of the peripheral continents to the mid-Indian Ridge are not so simple as in the North Atlantic. The continental shelf of east and southeast Africa is one of the narrowest in the world and is a product not of tensional faulting but of dextral transcurrent shearing along the Agulhas fracture zone. The zone itself is marked by a narrow trough (containing 2 km thickness of sediment at the base of the steep continental slope).

Nor does the western Indian Ocean with its many islands agree with the concept of sea-floor spreading. Stratified marine sediments of the northwest Indian Ocean show remarkable absence of distortion, perhaps because of the early removal of Arabia and Iran to their attachment to Laurasia (Fig. 3). Madagascar is geologically a fragment of Africa, withdrawn from alongside Moçambique (Flores, 1970) and carried northeast by further, similar transcurrent displacement in Cretaceous time (note the Moçambique, Prince Edward, and Madagascar fracture zones). The Mascarene (basalt) Plateau, bearing the granitic Seychelles in the north, doubtless travelled yet earlier, following

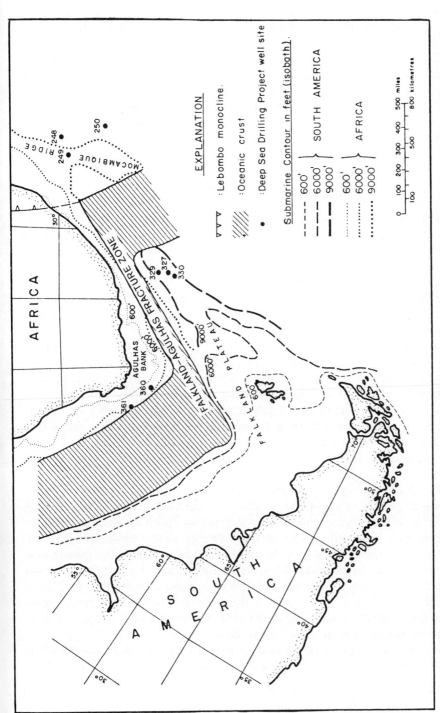

Figure 38 The opening up of the South Atlantic with the continental plates swinging apart upon the huge Falkland Islands–Agulhas fracture zone, which continues in the Indian Ocean off the coasts of Natal to Moçambique (after Rabinowitz and la Breque, 1979)

the Permian translation of Iran towards conjunction with Asia (Fig. 2(b) and p. 7).

The Carlsberg Ridge, northernmost sector of the mid-Indian Ridge, is relatively young. It is believed to have originated in the Gulf of Aden (the sides of which are fault-scarps) during mid- or late Cenozoic time (Wiseman and Sewell, 1937). This agrees with the modern Red Sea rift (p. 39), and with present seismic and rifted features of the ridge which testify to modern activity. Such a date, of course, precludes the ridge from association with continental drift, which was evident here first in Permian time with the northward withdrawal first of Arabia then of Iran–Afghanistan and opening of a southward gulf to Madagascar; and further opening up of Gondwana in mid-Jurassic–early Cretaceous times.

The Carlsberg Ridge is low and flat, and this remains a characteristic of the mid-Indian Ridge for most of its length. Precambrian metamorphic rocks have been dredged from both. The Indian Ridge is seismically and volcanically active in narrow belts, and has a crestal rift, yet its bathmetry and low level of present activity proclaim its probable youth as a simple cymatogenic rise. Vinogradov and his collaborators (1969) assessed the structure of the Earth's crust in the rift zone as 'a mosaic of normal crustal blocks and upper mantle material squeezed upwards. The first were formed by zonal melting processes with the division of pyrolite into peridotite and gabbro–basalts. The second are residual peridotites squeezed out upwards in a crystalline state and subjected to intense dynamic metamorphism with extensive production of mylonite, and at higher levels subjected to varying degrees of serpentinization (which) occurs at a depth of 10–15 km and arises because of the degassing of the mantle.* From the amount of titania and silica, the degree of differentiation of the mantle under the Indian Ocean is less than under the continents.' Gravity anomalies indicate reduced mantle density beneath the rift zone.

The Earth blocks of the upper zone are arranged *en echelon* obliquely across the main direction of the ridge; and detailed studies by the Russian team upon the crustal and subcrustal types of Earth blocks show thicker sedimentary cover with greater conformity to the normal suboceanic structure upon the former. The 'mantle-derived blocks, however, exhibit greater seismicity, and are almost devoid of sediments. Their component rocks are ultrabasic in origin with reduced and variable density due to the presence of volatiles. . . . The seismic activity of these blocks is two or three times higher than the "crustal" blocks. The heat flow at the surface of this region is three to five times more than on the ocean floor.'

These observations typify a cymatogen, and we note particularly that its development implies only simple, *vertically expressed activities and phenomena*.

---

* The water for the serpentinization of the mantle was juvenile water liberated by de-gassing of the mantle.

The western Indian Ocean is no simple tensional basin as the North Atlantic is often claimed to be. It is heavily sheared.

McKenzie and Sclater (1971) reported that they could not locate any magnetic anomalies between anomaly 5 and anomaly 21 on the Indian Ocean floor. 'This corresponds to a time span between 55–10 m.y. on the Heirtzler scale' (Fig. 24). Other scientists (Davies *et al.*, 1974) have recorded an absence of Oligocene sediments from, at least, the western Indian Ocean floor. Was there no opening of the Indian Ocean during the early Cenozoic?

More recently (1979) Norton and Sclater have used anomalies 16, 22, 28, 34, and M1 in conjunction with the known fracture zones of the southwest Indian Ocean to derive a series of reconstructions of the area from the time of the break-up of Gondwana to the present.

From Rodriguez the mid-Indian Ridge trends southeastwardly, passing north of the islands of Amsterdam and St Paul until finally it continues from west to east between Australia and Antarctica to the Macquarie–Balleny Ridge which here marks the boundary between the Indian and the Pacific plates (p. 43).

Between Australia and Antarctica, paired polarity reversal stripes (Weissel and Hayes, 1971) date back between 40 and 50 Ma, which places the separation of these continents in the early Cenozoic, after opening of the Tasman Sea (80 to 60 Ma).

Independent northward drift by Australia may then have continued until the Miocene (20 Ma) when New Guinea thrust into and began the great Pliocene subduction of the Banda arc (Norvick, 1979) (Fig. 3(d)). Antarctica reached its present position 30 or more million years ago and the present polar glaciation began (Stevens, 1980). Though the reversal stripes are paired they are described as 'systematically asymmetric'; and there are great changes of seismicity *along* the ridge, both in intensity and breadth of the seismic zone. South of Australia, too, the topography of the ridge becomes 'extremely rugged and the ridge crest cannot be identified either from the magnetic or the topographic data'. This part is, moreover, almost aseismic; and there are geological misfits under the hypothesis of tensional and spreading origins. These are all signs that the plate boundaries are sheared along irregularly curving plate margins. Finally, there is no evidence for corresponding subduction zones along the margins of either Australia or Antarctica.

The northeastern Indian Ocean has few islands and virtually no submarine features of proved continental origin. Until the Cenozoic northward drift of Australia it was part of the Pacific. The most important feature of this region is the Ninety-East Ridge (a gigantic megashear with associated volcanic structures) that trends due north–south from the Bay of Bengal to join with the main ridge beyond latitude 30°S. It is described as progressively younger towards the south, and a few seismic centres have been recorded along it (cf. Walvis Ridge); but these are insufficient to define its origin. Ninety-East is isostatically compensated, and Carey (1976) thinks that 'simple elevation of the isotherms, through out-gassing, causing phase changes, may satisfy all the (associated) geophysical data, as in many other ocean ridges (e.g. Shatsky, Eauripik-New Guinea, Ontong-Java).' Wilson explained it as a hot spot track adjacent to a major fracture zone.

In Asia beyond the orogenic zone, the Ninety-East megashear continues northeast as a series of fractures as far as the Kolyma, separating in China the region of interior Asian geology from the peri-Pacific structures of eastern China (Chen Gouda *et al.*, 1975). 'The Ninety-East' is truly a global tectonic phenomenon.

Sediments in the associated oceanic basins of the western Indian Ocean agree, in general, with the coastal plain and offshore sediments of East Africa in dating the local break-up as early Cretaceous. In the Moçambique and Crozet basins, however, a very deep borehole with sea floor around 5100 m and olivine basalt basement at 5800 m had a curious division of its 700 m core — 600 m was Miocene and Pliocene, which may be taken as the time of basin subsidence — by late Cenozoic vertical tectonics. The Cretaceous–Miocene unconformity is regional, yet throughout its thickness the Miocene includes broken Eocene and Oligocene materials.

In the same sector, joining the Crozets to the mid-Indian Ridge, the sheared southwest Indian Ridge rises to heights above the sea floor far in excess of the cymatogenic mid-Indian Ridge; also we have already noted that the length of the supposedly continuous submarine cymatogenic ridge from the equator in the Atlantic to the equator in the Indian Ocean measures almost exactly *twice* the length of the related African coastline. Like the Ninety-East megashear this sheared ridge is strongly seismic, and is reputed to be the scene of current volcanic activity. This ridge has often in the past been described as 'part of the world-girdling mid-ocean ridge system joining the Atlantic and Indian Ridges'. But this it is not: it has no crestal rift, and it is split longitudinally by enormous northeast–southwest dextral fracture zones. Moreover, there is no topographic connection with the mid-Atlantic Ridge for a gap exists between Bouvet and Prince Edward Islands (p. 4). But the final answer is given on p. 176.

In summary, the western half of the Indian Ocean basin is occupied by a hotchpotch of sea-floor elevations. Some of these reveal origins in Gondwanaland but little activity since the primary disruption; other elevations and basins display a full development of mid-Cenozoic to modern activity without any ancestral information.

In Palaeozoic geography this region was part of the proto-Pacific and the eastern Tethys, and it became a separate ocean only when the northern drift of Australia and New Guinea to junction with Indonesia shut it off from the Pacific realm. To this day its oldest bottom sediments are (like elsewhere in the western Pacific) Jurassic and early Cretaceous in age, as shown by nannofossils in DSDP boreholes 260 and 261, drilled into the abyssal plains of the Wharton basin which has been deep water since the mid-Albian when its location was more southerly. Early Cretaceous sedimentation rates are high; but Cenozoic deposits are thin.

The same history extends to the Naturaliste Plateau off Perth; and formations of similar ages transgress on to the Canning basin of northwestern Australia.

The nearby Java Trench transgresses, and is younger than, all these features.

The Pacific Ocean

The North and South Atlantic, Arctic, and Indian Ocean basins are all similar in type and origin. They are disjunctive basins formed during the late Mesozoic when the fragments of Gondwana and Laurasia drew apart, due to crustal expansion. They had not existed before. Their margins are fractured and monoclinal with a seaward dip of strata and of planation surfaces (p. 187). This dip was intermittently steepened about hingelines situated near the present coasts, so that the lands were uplifted and the sea basins deepened at intervals during the Cenozoic (p. 165). The typical landscapes formed by Cenozoic denudation were stepped planation surfaces (p. 187), and the continental margins are often fringed for long distances with the marine formations of emergent coastal plains. These coasts are aseismic and nonvolcanic.

The Pacific is quite different. Not only is it larger and deeper than all the other oceans combined; but it is patently older than they are for it was part of the pan-thalassa of Palaeozoic time through which Gondwana and Laurasia drifted majestically westward, and when these land masses collided and broke up, the proto-Pacific was the region into which all the daughter continents migrated. (There was nowhere else to go!)

The margins of the two parent supercontinents had been long established (evidently from the Proterozoic), and they had experienced a number of Palaeozoic and Mesozoic orogenies which cumulatively built a circumvallation of mountain ranges about each parent. After the fragmentation occurred, each of the daughter continents bore upon its advancing edge a relevant portion of the circumvallation (e.g. the proto-Andes, the proto-cordillera of western North America, the mountain systems of eastern Australia and of West Antarctica and so forth) which by conjunction now form the ring of mountains surrounding the Pacific. In addition, where the twin girdles of Laurasia and Gondwana met and joined across the Tethys Strait they now make the double-chains of mountain ranges stretching from Gibraltar to Burma (Mitchell, 1981). Herein the structures have been welded back-to-back, as on opposite sides of the granite core between the Himalayas (folded south) and the Kailas Ranges (folded north); or they make contact along large discontinuous structures like the Insubrian Line of Yugoslavia. Of the Mesozoic Tethys Strait practically nothing now remains.

All these orogenic belts were, by inheritance, highly seismic, orogenic, and volcanic; and these same characteristics (often with violent vertical tectonics) continued along them through Cenozoic time, uplifting new mountain ranges upon the old sites, and creating new marginal basins and island arcs offshore. All these environments, too, are highly seismic and strongly volcanic.

The Pacific is unique: it had its own origins and history in Precambrian time and should not be regarded as though it was simply a disjunctive basin of later geological time like the North Atlantic. Yet, curiously, the oldest sedimentary series known in the sea bed are early and middle Jurassic deposits in the basins of the northwest, near the Marianas Trench, and in the Wharton Basin northwest of Australia which belonged to the Pacific until the Miocene when Australian northward drift belatedly closed the gap with Indonesia, and opened the Woodlark basin east of New Guinea. These deposits are not much older than the Cretaceous formations of the other (disjunctive) oceans.

This chronology, and the existing state of the Pacific basin, may well have begun with a wide spread of lower Jurassic basalts over the only sea floor then existing. These basalts are found upon virtually all the southern continents (Drakensberg, Rajmahal) where they are sometimes nearly a mile thick. Beneath the sea floor only the top few metres of these Jurassic basalts have been penetrated by drill. Any marine sequence beneath them to correspond in time with the Gondwana–Karroo terrestrial group is therefore not yet known. Palaeomagnetic evidence by McElhinny et al. (1974) showed that Malaya was not a part of Gondwana.

The Pacific also seems to have participated in the second phase of plateau basalt emission and crustal activity found upon the lands (southwest Africa, Deccan of India and Serra Geral of Brazil) between 110 and 85 million years ago. Geomagnetic signals of this period are weak, and are therefore thought to have been originally of equatorial distribution, and the fact that they are now distributed in a broad swathe of geomagnetic quiet from Panama to Kamchatka has been interpreted to indicate northwesterly motion of the western Pacific sea floor from Cretaceous time onward. Interpretation of the regional sedimentation agrees with this (p. 153). Using orthodox views of sea-floor spreading and plate tectonics, first Larson and Chase (1972) and then Heezen and MacGregor (1973) and lastly Larson (1976) have explained the evolution of the western Pacific (beginning 124± 12 Ma ago) as involving a late Cretaceous to Miocene drift of the ocean floor towards the north-northwest. This accords in time and travel with the opening of the Tasman Sea and the independent drift of Australia–New Guinea to collision with the Banda arc which we have already noted (p. 153). Shallow-water upper Cretaceous is dredged from the tops of many guyots in the mid-Pacific mountains; but the model of plate tectonics becomes complex in the deep Pacific. There is evidence which has been interpreted as due to numerous former plates, now disappeared. Anyway it agrees with the general complexity and the age of the Pacific marginal lands, e.g. South America and Antarctica on both of which large vertical displacements in modern times have resuscitated tectonic features as old even as Precambrian.

The late Cenozoic and Recent tectonics of the North Pacific are dominated by the motions of the Pacific floor and its surrounding crustal plates. Slip vectors strike northwesterly parallel with the coast of North America, e.g. the coastal sliver from Vancouver, to Alaska (Yole and Irving, 1980; York, 1982) which slid from California; and later Baja California, which travelled northwest from Mexico along the San Andreas fault. The vectors become dip slip as the coastline changes direction off Alaska and the eastern Aleutians; become strike slip again off the western Aleutians and dip slip once more at the northern island arcs of Asia from Kamchatka to Japan (see Fig. 38).

Matching-up of magnetic anomalies of various slabs of the ocean floor, has shown, indeed, that huge areas of Pacific sea floor appear to have moved laterally by thousands of kilometres along very large faultlines. On the Mendocino fracture, for instance, shear displacement has been measured as 1160 km. The Pacific basin has enlarged enormously in post-Jurassic time.

Even at the present state of knowledge it is clear that the Pacific crust cannot have operated by simple sea-floor spreading in the manner claimed by sundry authors for the North Atlantic. It could operate in concert with other ocean floors only if all of them operated vertically, with sea-floor spreading but in conformity with an expanding Earth.

Consider the Antarctic–East Pacific Ridge, first researched by Menard (1964). This is not median to the ocean and is so askew in the Pacific as to be overridden in places by the North and Central American continent. Its heat flow, though augmented in places, is irregularly distributed and likewise the crestal rift valley is discontinuous. Near Easter Island the ridge spreads out into the enormous (Albatross) submarine plateau which stands in isostatic equilibrium. Where does the Antarctic–East Pacific fit into any global pattern of sea-floor spreading? And what ocean opens up to the east of it? Sediment types and sequences on both sides of the ridge are nonetheless similar as they would be if it was cymatogenically uplifted, despite the geographic differences to either side. Indeed the rates of sea-floor spreading to either side are not only the same, but are the most rapid of all the great oceanic ridges (10 cm/year). It appears to be a youthful feature, unrelated to the age and long history of the Pacific basin. The western edge of the rise is shown by Menard as passing a little east of the Hawaiian Islands, and reaching the Aleutian Trench about the longitude of Dutch Harbour. Studies by Herron (1972), confirmed by Hey et al. (1977), suggest that the present East Pacific Ridge was initiated at 55°S sometime between 50 and 65 Ma ago. The Chile Ridge part has been reactivated, with a crestal rift, during the past 20 Ma. There are modern volcanoes along some parts, but other sections are drowned, like the Emperor seamounts. The northern sector of the East Pacific Ridge came into existence only 10 Ma ago.

What then is overridden by North America, and comes into view again as the Juan de Fuca Ridge west of Canada? Or has a ridge feature risen under North America *after* that continent advanced to its present position? Support exists for both views in this conflict of continent and sea-floor ridge. Factually, where the ridge-form sinks from sight beneath the continent its surface zone of earthquake shocks, sporadic volcanism and even rift valley structures continues above a diagnostic substructure of levitated mantle (seismic velocity 7.2–7.9 km/s). This substructure may also be responsible for the Miocene 'plateau basalts' of the Snake River–Columbia region. For the present it may be regarded (with the Geological Map of the Pacific Ocean (Heezen and Fornari) before one) as a broad cymatogenic arch situated upon a gigantic megashear with westward dip passing through the points 45°N 143°W, 0° 102°W, and 67°S 180°W. This arch has grown progressively from south to north during the past 55 to 5 Ma. Transversely across this arch run numerous faults and fractures that are conjugate with the main structure (Fig. 42). Some of these conjugate fractures appear to enter the South American continent; but most are now tectonically inactive. With the thin oceanic crust, which favours development of these fractures, goes the relative suppression of a longitudinal crest rift zone. The cymatogen should *not* be viewed as a tensional 'spreading ridge' though there is marked symmetry of geological sedimentary formations to either side of the arch.

A much-researched area of the Pacific is the northeast (Vine, 1966; Morgan, 1968; McKenzie and Morgan, 1969; Atwater, 1970). During Palaeozoic time western North America was a marginal part of Laurasia and it grew by a number of orogenic cycles (Eardley, 1962). When Laurasia split and the North Atlantic opened, the western side of America, rotating about a Eulerian pole in the Yukon, is deemed to have advanced into the proto-Pacific with development of a trench and island arc. This movement ceased early in the Cenozoic and was succeeded (after an interval) by orogenesis. By the end of the Pliocene and during the Pleistocene the filled trench was elevated into the Coast Ranges to which the Cascades made the volcanic counterpart. The continent, we have been told, then travelled about 350 km from northwest to southeast. For these newest tectonics the San Andreas fault system has acted as part of the boundary between the American and Pacific plates, and expresses their relative motion during the past 5 Ma. That the San Andreas should be a simple dislocation between two rigid plates is unlikely. Displacements are known on other faults; the Idaho disturbances and the Basin and Range fault systems must be considered; and from Carey (1978) comes the idea that western North America is a very wide, soft boundary between two rigid moving plates. This seems to be near the truth.

On the opposite side of the Pacific Ocean the Asian plate is deemed to be overriding the Philippine Sea plate as the latter is consumed by subduction at the Nankai Trench off Shikoku (Fitch and Scholz, 1971). Modern earthquakes (e.g. Nankaido, 21 December 1946, magnitude 8.2) have been quoted in support. But these are local. To suggest that sea-floor trenches associated with the Mesozoic marginal basins of eastern Asia are the subduction sinks related to the East Pacific Ridge (much younger and the full width of the North Pacific Ocean away), as the Atlantic-derived hypothesis of sea-floor spreading suggests, is absurd.

Notwithstanding, the following stratigraphic events of the *northwestern* Pacific call for correlation with events in northwestern America:

(a) 'A middle Eocene to late Oligocene turbidite sequence from Alaska occurs beneath the Aleutian abyssal plain, with small variety of Eocene species, only small movement of the plate in high latitudes at that time.' (*Geotimes*, 1971). The area was near the Yukon Eulerian Pole so displacement was small.

(b) Farther to the west only pelagic rock types appear. Towards Kamchatka very thick, Miocene pelagic clays covered a large area of the sea floor.

(c) Three million years ago volcanic effusions succeeded over the whole region from the eastern Aleutians to Kamchatka and the Kurile Islands. These agree in time with the opening of the Gulf of California and much conflict of movement between the Pacific and American plates. They show that the tectonic development of the northern Pacific had then nearly reached its present stage. These three varied and distant effects show how far-ranging and complex a single related series of plate conflicts may be.

Meanwhile, Krause (1971) has summarized the characteristics of the Celebes Sea–Sula Sea region:

(a) the Philippine fault may have at least 110 km of horizontal displacement;
(b) there was no land in the vicinity prior to the Cretaceous;
(c) much spilitic lava was extruded during the late Cretaceous and early Cenozoic; and
(d) there was major orogeny in late Miocene time, and another orogeny, still active, began in late Pliocene and Quaternary time.

A useful evolutionary synthesis for the western Pacific region and its margin was put forward in plate tectonics terminology by Hilde, Uyeda, and Kroenke (1977). This is an example of how plate tectonics thinking can simplify a naturally tangled set of data; but we cannot expound it in detail here, and the authors intend to continue the study.

Magnetic polarity data have been taken by geophysicists to indicate that sea-floor spreading has been active on the Indian–Antarctic–East Pacific Ridge, but again the inference of crustal movements is dubious and there are no corresponding subduction zones about West Antarctica.

The overall pattern of sedimentation in the South Pacific seems to reflect modern bottom currents, with changes of the Antarctic convergence in the southern ocean, rather than record any progressive displacement of the oceanic crust.

A striking feature of the Pacific basin is the close correlation in some sectors between gravity anomalies and topography. Not only are large negative anomalies associated with sea floor trenches, but as the trenches are approached from seaward the sea floor rises before plunging downwards and so a border of positive anomalies (of about ±50 milligals) is consistently recorded. So the North Pacific is ringed about by positive as well as negative anomaly belts. These are believed to be due to elastic bending of the crust under load (cf. Gunn, 1949).

Conclusions from the Great Southern Ocean

The Great Southern Ocean is likewise a large area of open ocean wherein comparative and synthetic studies may be made of the several oceanic basins and ridges which meet there.

The first step in the creation of the Great Southern Ocean was taken in the Cretaceous (anomaly M11) when South America separated from Antarctica at Drake Strait. The Scotia arc did not develop until 30 Ma ago.

Meantime, the most important step was when Australia separated from Antarctica and drifted north to collide with southeast Asia (about 55 Ma ago). This opened the way for the West Wind drift to sweep from Cape Horn right round the southern hemisphere between latitudes 50°S and 60°S back to Cape Horn. There was henceforth no intervening continental mass and no hindrance to marine life which was swept onward to make a great uniformity of fauna and flora in this circumpolar zone. A normal west–east mid-ocean ridge was developed between the two continents (beginning 55 Ma ago); but no subduction zones or trenches exist along either the Australian or the Antarctic coast. Indeed, even today the geometric fit between Australia and Antarctica (taken at the edge of the continental shelves) is remarkably good.

We have already seen that the South Atlantic Ridge ceases near Bouvet Island and does not continue to the mid-Indian Ridge. The connection between them is made by the long shear ridges of the southwest Indian Ocean, of quite different structure and significance, though they may act (synchronously or separately) as boundaries under plate tectonics.

The mid-Indian cymatogenic Ridge is also not directly continuous with the Antarctic Ridge. There is a gap followed by an area of rough sea-floor topography between them south of Australia, and the plate boundary between the respective ocean basins is formed by another ridge at right angles to them both (the Macquarie Ridge). This gives a curious relationship, for the Tonga–Kermadec arc shows *underthrusting* by the Pacific plate whereas the same plate is indicated as being *overthrust* on the Macquarie Ridge. The Alpine Fault of New Zealand emerges as a joining link between two regions of contrasting dips.

If this seems questionable, we may note that the Philippine fault likewise 'appears to connect a zone of underthrusting of the Pacific floor near the Philippine Trench with a region of overthrusting (by that same floor) west of the island of Luzon near the Manila Trench'. (Isacks, Oliver, and Sykes, 1969)

The mean depth to acoustic basement is apparently greater on the east than on the west side of the Macquarie Ridge. Also the geomagnetic reversal stripes of the Indian and Pacific Oceans on opposite sides of that boundary (Fig. 39) are independent of each other in direction and in dating. Nor are the paired reversal stripes parallel with the Macquarie Ridge, though the geomagnetic stripes are symmetrical on both sides up to number 19, which can be taken to be the date (50 Ma) of separation (Eocene) of the Australian and Antarctic continents (Weissel and Hayes, 1971). The cross-section of the ridge is here very wide, with many cross-fractures. These conjugate shears began to develop about the mid-Miocene (20 Ma ago) and continue to operate.

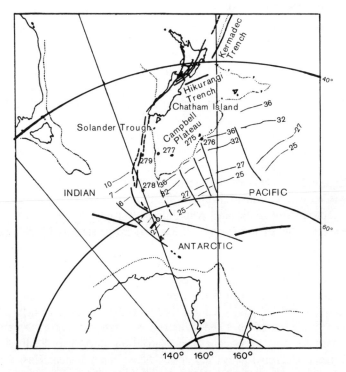

Figure 39   Triple junction south of New Zealand between the Indian, Antarctic, and Pacific plates. Bold lines are megashear ridges, medium lines are conjugate shears, light lines are geomagnetic isochrons. Numbers 275–279 are DSDP leg 29 sites (*Geotimes*, 1973). Showing lack of agreement between data from Indian Ocean and Pacific Ocean (after Katz, 1974)

The Macquarie–Balleny Ridge has been variously interpreted as a former island arc, a shear zone, or a mid-ocean ridge marking crustal plate separation. Present evidence on currently active tectonic mechanisms is inconclusive; but transverse profiles, topography, gravity, and magnetic data are all in clear accord with the hypothesis of a shear boundary. Connection with the New Zealand Alpine megashear also seems probable. Marked uplift, however, is involved and Macquarie Island itself has been interpreted as an upcast piece of Pliocene oceanic crust* bordered upon *both* sides by trench-like depressions with gravity anomalies of −150 to −200 milligals and containing little sediment. At intervals the trenches are linked by deep, narrow, transverse depressions through the ridge.

Our guide in this region is H. R. Katz (1974) who has assembled much data from DSDP and oil exploration surveys, and concluded that in the southwest Pacific various segments of crust and different types of continental margins are juxtaposed. They differ 'in age, structure and tectonic regime'. The region of interest lies between Antarctica and the Tonga Islands, and 'extends roughly along the 180th meridian'. 'The continental margins referred to border the Pacific Basin proper . . . but are not synonymous with plate boundaries as defined by plate tectonics theory.' They may coincide, however, as along the Tonga–Kermadec ridge–trench system; but in the New Zealand region they are entirely within the Pacific plate. Here the Pacific–Indian plate boundary is assumed to pass through the middle of a continental block (le Pichon, 1968), while south of New Zealand this boundary separates oceanic crust on both sides (Fig. 39). Katz notes that New Zealand is only part of a much larger continental block which includes the Chatham Rise (east) and Campbell Plateau (south). The whole 'is part of New Zealand by virtue of both structure and geologic history. The crustal thickness is, however, considerably reduced' (17–23 km). The area is isostatically and tectonically stable and is mainly aseismic.

The western edge of the Campbell Plateau drops steeply by normal faults to the Solander Trough (2.5–3 km of turbidite infill) and Emerald Basin (3–4 km deep), the two being 1100 km long. This depression continues into New Zealand as the Waiau Cenozoic basin. West again is the Macquarie–Balleny Ridge, a shear zone which projects into the New Zealand Alpine fault, and the composition of which is considered entirely oceanic. Early Miocene sediments are associated, but the structure is mostly volcanic (tholeiitic).

Noteworthy is the view that the plate boundary represented by the Tonga–Kermadec Trench (Fig. 39) enters New Zealand near East Cape and after passing through both islands as a series of well-known major transcurrent faults, departs south-southwest from Fiordland to Macquarie and there bends southeast towards the ice-capped Balleny Islands before being lost off Antarctica. The system has been active for a long time, movements are known of Cretaceous, Miocene, Pleistocene, and Recent date in New Zealand (Stevens, 1980).

In the past the Hikurangi Trough has been compared with the Kermadec Trench; but this no longer appears to be valid. The Hikurangi Trench has been found to differ in several important aspects. The related Benioff zone

---

* Shown at DSDP site 279 where lower Miocene sediments immediately overlie basement of vesicular basalt.

of Kermadec does intersect the surface of the Earth close to the axis of the negative gravity anomaly; but this intersection lies in the middle of the North Island continental crust and 200 km west of the Hikurangi Trench (Hamilton and Gale, 1968). This trench can no longer be regarded as a continuation of the Kermadec Trench and as Hatherton (1970) and later Katz (1974) have pointed out: 'if the course of (modern) shallow earthquakes above the Benioff zone is sought in terms of overthrusting of lithospheric plates . . . no evidence of (contemporary) underthrusting of the eastern half of the continental crust, along the Benioff zone in New Zealand, is found from geological information. . . . The trench sediments, probably Plio-Pleistocene in age, gradually thin to the northeast, while in the same direction the trench becomes wider and flatter' until structurally and morphologically it dies out.

Tentatively, it may be suggested that the Hikurangi Trench is related to the kinking which affects all the transcurrent faults in the vicinity of Cook Strait and that it is related to the Kaikoura fold system in which similar elongate basins were repeatedly formed during the late Miocene and Pliocene and were partly filled with thick turbidite sequences (Kingma). Mesozoic basement is also known on both sides of, and beneath, the Hikurangi Trench, showing the presence of continental-type crust.

Also involved with these problems is the origin and chronology of the Tasman Sea. In Mesozoic Gondwanaland, New Zealand was the marginal part of Australia (Fig. 40) and the two travelled eastward together following the disruption of that supercontinent. Their separation from each other came later, beginning in the south during the late Cretaceous according to the sea-floor drilling data, and the separation was completed *via* the Coral Sea to New Guinea by the Eocene. New Zealand appears to have remained static in its present position while Australia and New Guinea undertook a new drift to the northwest (away from the Pacific) and driving into the Banda arc during the Miocene (according to van Bemmelen), separated the Wharton Basin (with Jurassic oldest rocks) from the rest of the Jurassic basins of eastern Asia in so doing. A general drift north-northwest is manifest also (p. 136) in the sea crust of the western Pacific as far as the Emperor seamounts, from the late Cretaceous until the end of the Miocene. At the same time this part of the ocean floor underwent regular subsidence.

At about the same time (50–60 Ma ago) Antarctica separated from Australia and moved south to its present polar position, arriving there closely about 20 Ma ago, when the continent became even more heavily glaciated than it is now.

Evolution of the Tasman Sea–Coral Sea basin (see Fig. 40) does not embrace sea-floor spreading from a mid-ocean ridge — there is none; but two long ridges rising from the sea floor are, by geological samples, of continental origin allied: (a) to Australia (Lord Howe Ridge); and (b) to New Zealand (New Caledonia). Their geologic affinities are Palaeozoic and early Mesozoic respectively. The whole area of the North Tasman–Coral Sea is, indeed, underlain by continental crust, described as 'thinner under basins'

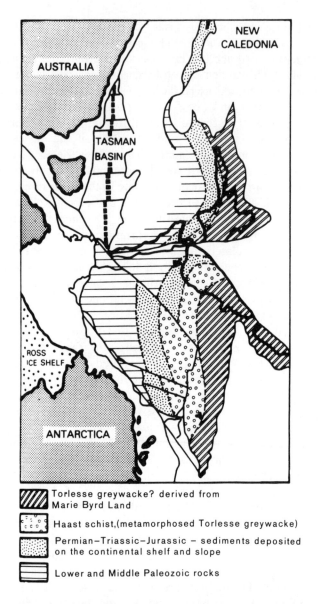

Torlesse greywacke? derived from
Marie Byrd Land

Haast schist,(metamorphosed Torlesse greywacke)

Permian–Triassic–Jurassic – sediments deposited
on the continental shelf and slope

Lower and Middle Paleozoic rocks

Figure 40　Reconstruction of southeast Australia, New
Zealand, and adjacent regions of Antarctica, at the time
the Tasman sea opened (after Stevens, 1979)

and 'thickened under the Lord Howe Ridge'. There is also a noteworthy absence of late Paleocene–early Eocene and late Eocene–Oligocene sediments which is perhaps correlative with the smooth planation of Australia at those times. A view of the bauxite workings along the Gulf of Carpenteria, with the deposits capping marine-cut cliffs which are smoothly topped by the great Australian 'moorland' planation (early Cenozoic) surely convinces the beholder that, at the time the bauxite cap was forming across the smooth (early Cenozoic) landscape the quantity of denudational waste shed into the sea from Australia must have been almost nil.

The story of New Zealand, amid these various tectonic influences, is told by the pen of Dr Graeme Stevens (1980) in *New Zealand Adrift*. He has graciously granted permission for the reproduction here (Fig. 40) of the region about New Zealand at the time when the Tasman Sea began to open. Note that the Torlesse rocks, and their equivalents the Haast schists, are believed to have originated from Marie Byrd Land in West Antarctica!

About the region of Tierra del Fuego there is no connection between the East Pacific and South Atlantic Ridges. Geomagnetic stripes between the two oceans are difficult to reconcile and are perhaps best explained on the geomagnetic 'shadow' hypothesis involving only vertical movements in the sea floor, which renders a better picture of the bathymetry and flank sediments of the Atlantic and Pacific Ridges than does lateral sea-floor spreading. The age of Weddell basin is discussed by J. L. La Brecque and P. Barker who note that it occupies the space between the East Antarctic craton and the post-Palaeozoic orogenic belt of West Antarctica, and is related to anomalies 33 and 34 and also M29–13. High heat flow values in some parts are indicative of late Cenozoic movements.

In the region west of Cape Horn and in the Bellingshausen Sea Craddock and Hollister (1976) sank four boreholes on DSDP leg 35; but obtained no oceanic crust older than Cretaceous. Typical of the southeast Pacific all the sites yielded Pliocene sediments with local Eocene, Oligocene, and Miocene deposits; Miocene was deep water glacial. The Bellingshausen Abyssal Plain lay at 4560 m depth. To be accounted for, moreover, is the Scotia arc, the northern half of which is tipped in mid-ocean by South Georgia. This island, which has recently been compared geologically with Southern Patagonia (Dalziell and Elliot, 1971), is composed of Palaeozoic and Mesozoic geosynclinal rocks now superbly overfolded towards the north. Aseismic and nonvolcanic, the island was formerly an integral part of the South American cordillera. Separation was by strike-slip displacement of about 1500 km. The southern (South Orkney) branch of the Scotia arc is of strongly folded earlier Palaeozoic rock systems and it roots back in the Antarctic Peninsula. Together the two lines manifestly result from the continental separation of South America from West Antarctica and was a late Mesozoic event. At their eastern extremities the two features are now united by the modern arc of South Sandwich: barren, smoking volcanic islands built upon the seismic zone of the accompanying deep-sea trench of the same name. All are currently active.

Dalziell and Elliot concluded 'the structural framework of the Scotia Arc in terms of plate tectonics is not entirely clear'; and from my personal inspections of all three parts of that arc (King, 1969) I would have heartily agreed, especially if the arc were to be associated with any simple hypothesis of progressive sea-floor spreading.

But in 1977 de Wit carried the study further, and gave a most admirable synthesis of the Scotia arc (Fig. 41) in terms of plate tectonics, dating its origin from the division of South America from Africa in the earliest Cretaceous when this part of the Gondwanaland circumvallation in some ways appears to have been fringed with marginal seas like those now existent in eastern Asia. With the Cenozoic history interpreted by Dalziell and Elliot

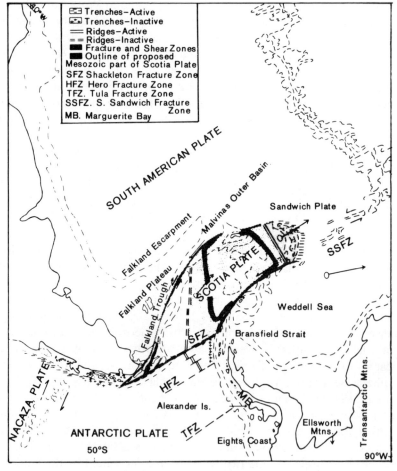

Figure 41   Plate tectonics of the Scotia Arc (after M. J. de Wit, 1977)

(1971) a picture is obtained of the South American plate moving left-laterally with respect to the Antarctic plate and a 'complex zone of micro-plates beneath the Drake Passage and Scotia Sea'. 'Nevertheless' de Wit concedes, 'the boundary between the American and Antarctic plates remains one of the least understood major plate boundaries in the world.' (Fig. 41)

Since de Wit, the only major contribution to the geology of this remote area has been the thematic set of papers on West Antarctica and the southern Andes published in the *Journal of the Geological Society of London* (1982) which appeared as Vol. 139, part 6 while this book was being published.

Ocean basin chronology

The student of submarine phenomena in austral oceanic basins finds himself beset by a North Atlantic group of ruling theories that often seem inadequate in the southern oceans. Phenomena are so much more complex, and when he reaches south to the Great Southern Ocean there is added the need to reconcile tectonic data as between the various ocean basins into a single coherent system. This requires more than simple sea-floor spreading, and even plate tectonics has not yet solved all the problems of the junctions between the basins.

In the south, where Atlantic, Indian, and Pacific Oceans unite in one great storm-tossed ocean girdling the globe between latitudes 50°S and 65°S, at least two and probably three stages of development need to be postulated to explain the bathymetry and tectonics of the sea floor (cf. Zhivago, 1971). A simple sea-floor spreading beginning in the Cretaceous and continuing more or less uniformly through the Cenozoic is inadequate. The only conclusions possible are that:

(a) like cymatogens, rift valleys, and basins upon the lands, the suboceanic ridges are composite features embodying parts of different origins and ages all united since late Cenozoic time in a set of Pliocene to Recent *vertical uplifts*;

(b) many parts bear no relation at all to large-scale horizontal Earth movements originating during the late Mesozoic.

Chronologically the three stages at present recognizable may be classified:

(a) Probably Palaeozoic relicts of former suboceanic bathymetry such as the trans-Pacific lineament from northwestern Chile to the Emperor seamounts. These relate to the Pacific only (because of their age: there were no other oceanic basins at the time). Most of the features on this lineament are now dead (Darwin Rise, Albatross Plateau); though others still have a propensity to volcanic and seismic activity, showing that they still function as plate boundaries (e.g. Hawaiian chain).

(b) Features originating as a result of the late Mesozoic break-up and dispersal of Gondwanaland, the continents, minicontinents, Scotia arc, and parts of the suboceanic ridges. These display continental-type composition and structure (e.g. Kerguelen), and completed, or nearly completed, their continental drift before the end of the Mesozoic. Some have been rejuvenated by the later effects of stage (c). Insofar as geologic evidence within the Pacific goes, it favours not the slow, regular horizontal displacement throughout Cenozoic time of the thermal convection hypothesis; but supports convergence of the continents into the proto-Pacific, completed shortly after the end of the Cretaceous, followed by long Cenozoic quiescence and much freedom from horizontal displacement.

(c) Beginning early in the late Oligocene or early Miocene, and increasing greatly to a climax in Plio-Pleistocene time, new *vertical*, cymatogenic movements generated arches and domes and ridges and crestal rifts upon the floors of all the major oceans. These affected both the features formed during the Cretaceous disruption (or earlier) and entirely fresh areas of sea floor. The result was the present mixed pattern of active and non-active submarine features in the Pacific and other southern oceans. Further phases of this development were again active during the Pliocene and Pleistocene, and in some areas continue with lesser intensity to the present day.

Such an interpretation involving three stages of development within the ocean basins is entirely consonant not only with sedimentation in the ocean basins, but with the geomorphic evolution of the land areas of the globe (King, 1962/7, 1976) (p. 187), and is almost certainly a consequence of similar types of tectonics on both (Fig. 50). Mostly the controlling factor is vertical uplift of certain zones, with renewed subsidence in ocean basins. The apparent horizontal displacements which are not plainly associated with transcurrent faulting, find explanation in radial movements upon a sphere.

Volcanism in the ocean basins

In almost every ocean the sea floor is peppered with volcanic cones. Most of these lie dead within the depths of the ocean basins, but here and there an exceptionally active or long-lived cone breaks the ocean surface as a volcanic island. The Pacific Ocean basin especially is dotted with volcanic piles, about fourteen thousand of them. This abundance may be a function of the basin's greater age. Some are of immense size. Mauna Kea, Hawaii, rising 10,203 m from the Pacific floor, to reach 4000 m above sea level, is the biggest single mountain cluster on Earth.

Marine volcanic islands which have been truncated by the waves and since subsided below sea level are called guyots. Most of them seem to have sunk by 600 to 2000 m and it is evident that they afford a measure of the amount by which the ocean floor has sunk in later geologic time. The Pacific floor especially has subsided, and if this had happened in the second half of the Mesozoic, it could well be a cause contributing to the continental drift of

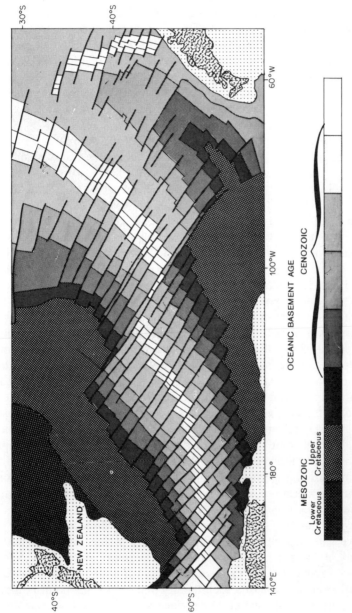

Figure 42 Conjugate cross-fractures of the Antarctic and East Pacific Ridges with the main direction of shearing along the ridge-axis. Compare with Fig. 14 (after Heezen and Fornari, 1976)

that time. Sclater and Dietrick (1971), following an initial analysis by Menard (1969), have gone further and shown that, apart from the North Atlantic between latitude 40°N and Iceland at 66°N, there is a uniform relation between depth and age of the oceanic crust. They demonstrated that 'the East Pacific Rise in the North Pacific shows a uniform decrease from a crestal elevation of 2800 m to a depth of 5600 m for 75 m.y. old crust. All other mid-ocean ridges (with the exception of the mid-Atlantic Ridge south of Iceland) show the same uniform increase of depth with age.' Joides drillings show three basins of unusual depth: (a) near Hawaii; (b) the southern Argentine basin; and (c) the Wharton basin northwest of Australia. In the first and last the oldest sediments are Jurassic.

It is to be expected that those upraised parts of the ocean bed which form the major suboceanic ridges should exhibit numbers of volcanic islands along their length. This they do, all of them, and the islands on the whole are very similar in composition: alkaline and olivenic basalts, phonolites and tinguaites, and some ultrabasic rock types. This small range of rock types may be taken to be of mantle origin.

But a few other land areas associated with the submarine ridge system (and agreeing with it in seismic, tectonic, and volcanic activity) show sedimentary and metamorphic rocks, and definitely continental foundation. Such are Albatross Plateau, Madagascar, New Zealand with the submarine Campbell Plateau, and western North America.

As volcanoes can be of any age in any part of the lands or ocean basins any polarity reversal may be met with in lavas anywhere — after the date of the local crustal flows. Reversals may succeed stratigraphically in consort with the flows. On the island of Reunion in the Indian Ocean, for instance, basalts up to 1 Ma (K–Ar measurement) with normal polarity overlie basalts of 1–2 Ma with reversed magnetism. To determine the age of the local crust only the oldest extrusive basalt counts! To check this, drilling is usually continued for several metres into what is, on probability, likely to be the true crustal basalt. But there is still doubt whether the early assumption: 'Magnetic susceptibility, remanent magnetism and Qn values obtained from the rocks indicate that permanent polarization is responsible for the geomagnetic anomalies measured at sea' is really true. As at present, wandering of these values through rock masses cannot be excluded.

A sufficient number of instances is quoted of Cretaceous basalt floor (radioactive date determination) overlain by Cretaceous sediments dated by fossils, and of Miocene basalts (radioactive dating) overlain by Miocene fossiliferous sediments. In each case the relationship is expectable and normal, either at the base of a sedimentary pile or at some point in the pile. What is necessary to demonstrate sea-floor spreading, say, throughout the Cenozoic is a succession of these relations transgressionally across the surface of the crust.

Where there is a discrepancy between the age assessed for the supposed oceanic basalt crust and the immediately overlying sediment, if the basalt be younger then it is a later intrusion and is not the true oceanic crust. If, on the other hand, it be markedly older than the basal sediment an interval of non-deposition, or of erosion, is to be assumed between the two. A close agreement of age between pillow lavas and the succeeding sediment is ideal, and affords the best test that the true crust has been reached by the test core.

The curious history of the Mediterranean basin

Compared with the major oceans the Mediterranean seems almost insignificant. But its situation is unique, and its history is complex and strangely enlightening. During Palaeozoic time Gondwanaland and Laurasia were separated from one another by an east–west (Tethys) sea or strait; but when the continents collided in the Mesozoic (p. 6) the Tethyan basin was crushed between them in a tectonic welter of Earth blocks (Dewey *et al.*, 1973). Some authors write of an anticlockwise rotation of Corsica and Sardinia; others associate 'the partly chaotic allochton of the Appennines' with 'subduction and submarine sliding on the advancing edge of the Earth block'. After the initial impingement in the west (eastern North America against northwest Africa and Spain, the Tethys remained open in the east (from western Europe to the Himalaya) and continued to function as a marine sedimentary trough throughout the Cretaceous and early Cenozoic.

To begin with, south of the Alpine zone was a kilometre-thick Mesozoic pelagic facies of carbonates that extended westward to the Bahamas and rested on a continental basement of granites and metamorphic rocks belonging to the continental shelf south of Tethys. From Africa the supply of clastics and organic matter was minimal.

On the Atlantic slope (DSDP legs 5 and 14) the records are of shelf deposits from an arid landscape: mudstone, clay, sand with nannoplankton ooze, or sometimes shale and silicified sandstone beginning with early Cretaceous. Mid-Cretaceous and upper Cretaceous follow. Then there is an hiatus (unconformity) of 50 Ma to mid-Tertiary nannoplankton chalk ooze. This hiatus surely corresponds with the planation, known all over Africa, of the 'Moorland' cycle of landscape development (Chapter 10) ranging from the late Cretaceous to the Miocene. Similar hiatuses were obtained in boreholes 135, 136, 139, and 140 in the eastern North Atlantic, and in the western North Atlantic (beyond the mid-Atlantic Ridge) from the records of sites 99, 101, 105, and 106.

Off Liberia (DSDP leg 5) 'The continental margin bears evidence of major structural elements that may be tied to similar events off South America illuminating drift.'

During the lower Jurassic, block faulting of this zone became manifest as the supercontinents approached collision. The Tethys deepened; and irregularity of the ocean floor increased, with more local variety of sedimentation during the middle and later Jurassic as a result of collision (p. 6). The western end of the Mediterranean basin was then closed, and a true Mediterranean Sea was begun. Cretaceous deposition therein greatly smoothed the sea floor, which deepened as the continental margin continued to founder. 'Vari-coloured marls with red and white coccolith limestones were widespread' (Bernoulli and Jenkyns, 1974), associated with ophiolites in the orogenic zone to the north. This Tethyan facies of the Alpine–Himalayan zone then reached from the Caribbean to Indonesia.

The final stages of the collision were marked by the Alpine orogeny during the Oligocene which closed off most of the eastern section and there united fragments of Gondwana (e.g. Iran, Afghanistan, and India) with Laurasia. The only part of the Tethys remaining was the Mediterranean, which towards the end of the Oligocene was almost landlocked.

The Tethyan region, including the Mediterranean, has been a tectonically active zone at intervals ever since, with peaks of activity at the end of the Oligocene, again at the end of the Miocene and yet once more at the end of the Pliocene. Each episode renewed the tectonic features of the landscape, and was followed by a quiet intermission during which stratigraphic sequences were laid down in local basins. These Cenozoic sequences sometimes contain alternations of marine and continental facies which help in detailed correlation of Cenozoic deposits from one district to another.

This later history of the Mediterranean basin, which is about 3000 m deep, is quite extraordinary and is reported by Hsu (1972) from *Glomar Challenger* results. With the tectonic movements at the end of the Oligocene the Mediterranean became entirely landlocked. To the north was another extensive water body (called by the French *Lac Mer*) which stretched unbroken from the Plain of Hungary past the Black Sea and Caspian Sea to far beyond the Aral Sea. It received cold waters from the north and at one time drained into the Mediterranean. But this supply was diverted by the rise of the Carpathian mountain system.

The Mediterranean is an area of strong evaporation; by the middle Miocene it was reduced to an empty desert basin dotted with playas in which strong brines deposited gypsum and anhydrite with at least one bored thick deposit of rock salt. That this former (and present) sea area should have been reduced by evaporation to a desert 3000 m *below* sea level during the Miocene seems unbelievable, but the distribution of evaporites in a pattern of desert lakes, the existence of high temperatures (anhydrites are precipit-ated only above 35 °C) and the occurrence of stromatolitic dolomite (which requires sunlight for its development, not the gloom of ocean depths), and the occurrence of mudcracks filled with wind-blown silt in some cores, is strong evidence. The Red Sea basin was similar. And finally, countries around the Mediterranean — France, Algeria, Libya, Syria, Israel, and others — have very puzzling gorges, cut through hard rocks and descending in the coastlands as much as 1000 m below sea level. Even the rivers which cut these Miocene gorges were insufficient to counteract the high evaporation rate of this appalling Miocene basin.

The Strait of Gibraltar was breached anew early in the Pliocene and the Atlantic Ocean poured in, as it still does. The gorges were then filled with new marine sediments (and new fossil faunas) of that age. Within a few thousand years the basin was again full to the brim and deep, soft marine oozes were again laid upon the Balearic abyssal plain. Three thousand metres thick, the Pliocene beds have been intruded massively by salt domes rising from the Miocene evaporites below.

Truly, this is a very special and informative history, unsuspected before the modern era of oceanographic research. Surprising as some of the episodes are, they nonetheless comply with the general datings of events in other regions, and conform to standard types of environment.

The orogenic framework of the Mediterranean basin is also of unusual interest, and crustal plates, both large and small, have been much studied. We cannot here summarize this fascinatingly detailed work. Interested readers will find the following four references rewarding, amongst others: Dewey and Bird (1970); Dewey, Pitman, and Ryan (1973); Alvarez (1973); Alvarez, Cocozza, and Wezel (1974).

Smaller marine relicts to the east are the Black Sea and the Caspian Sea, the southern part of which is underlain by oceanic-type crust. The further extension of it eastward is crushed between Laurasia to the north and Iran with India, the two parts of Gondwanaland which drifted from the south across the equator and united with Laurasia. A suggestion has been made that the Andaman Sea south of Burma, which otherwise appears anomalous, represents the farthest east portion of Tethys where it opened into the proto-Pacific Ocean.

The history of the Mediterranean is not unique. Sea basins are not infrequently locked in between colliding crustal plates, and other basins acquire salt deposits when they are newly opened to the sea. East Africa provides an example, with 3000 m of halite in the coastal Triassic sequence (Kent, 1972). During the Permian a gulf of the sea opened here from the north and left fossiliferous marine deposits not only on the edge of the African mainland but also in northern Madagascar. Iran, which according to its geology belonged at that time to Gondwanaland, became attached to Laurasia (perhaps by minor impingement before the major Jurassic collision of the supercontinents about northwest Africa (p. 7; Figs 3b, 3c)) and was whisked away to the north, locking in temporarily an arm of the sea as it went. During the Triassic this local sea dried up, thus producing the thick salt deposits of the East African coastlands.

The balance of Earth and Sea

The plenitude of postwar researches into oceanography has given rise to many theories regarding the evolution and bathymetry of the sea floor, and some changes of ocean level through the Cenozoic and Quaternary eras.

To begin with the Earth was about one-sixth of its present size. It was completely covered by 'continental crust', had no oceans at all, and was devoid of life. Much of the matter in space was of this type — dark, cold, and very dense.

How different is our Earth now, shining blue in colour from the abundance of present oceans (70% of the Earth's surface) and decked with white clouds constantly varying in pattern from the weather in the atmosphere (see cover). Only the continents show as remnants of the original crust that was torn asunder by expansive forces from within the proto-Earth; and then carried apart by continuation of the same forces acting radially within the sphere (see Fig. 31).

The specifications were:

(a) the development of light elements within the body of the planet;
(b) combination of those light elements in abundance at temperatures that were amenable for transport both within and without the Earth, perhaps within bodies of rock;
(c) the formation of an end-product stable at moderate temperatures;
(d) the action must be widely available, worldwide if possible.

In a word, **steam**, the activator of so much volcanism in and around the Earth, and whose end-product, after cooling, is water: hence the vast growth, through geological ages, of the oceans.

So the original dry planet was changed by addition of an atmosphere and hydrosphere; which it was large and dense enough to hold (under gravity) as outer envelopes. The materials were the steams and gases of volcanic eruption which, having burst asunder the continental crust of proto-Earth, condensed as water and accumulated as seas and oceans. During this process they caused the expansion which is characteristic of the mantle zone within the Earth.

An abundance of clouds brought about normal weather controlled by the water cycle of the atmosphere, ensuring non-polar temperatures over most of the Earth. How many planets have this temperature regime combined with an optimal distance from the Sun?

But problems nonetheless do exist. Rezonor (1978), for instance, after noting from the Indian Ocean that many drill holes yielded evidence of deep-water sediments and an absence of shallow-water phases* went on to review the deep-water sedimentary basins of all the major oceans and concluded: 'Thus, for all the oceans of the planet (except for the Northern Ocean, which has still not been studied from this aspect) major sinkings of the floor, measurable in kilometers, have been identified embracing at least half the water area they now occupy! This geological phenomenon is so huge that it has no equivalents on the continents. In part we may trace the sinkings since the second half of the Cretaceous epoch. There is no doubt that this process occurs in Cenozoic time.'

* Cf. the depo-centres of the Mississippi Delta described by E. G. Murray (King, 1962/7).

Let us glance again at Fig. 31. Therein the expansion of the southern hemisphere below the oceanic areas is concentrated in three great wedges — the Pacific, Indian, and Atlantic wedges — in each of which a huge sea-floor ridge has, in the last 5–6 million years, risen into existence. These rises are perhaps relics of the tongues of 'anomalous, volatile-charged mantle' responsible for the expanding phenomenon of the planet.

In the oceanic wedges, widening of the oceans between the continents becomes large indeed, and requires a constantly increasing amount of additional water from volcanic sources. There comes into play a fluctuating balance between the volume of new space required under expansion to meet the spherical geometry of the planet and the supply of new water from volcanic activity. The units of each, of course, are measured in cubic kilometres.

It is difficult to determine for these wedges what imbalances may result between the amount of expanding mantle material invading them from below and laterally, and the quantities of water filling in the oceans above.

Rezonor, writing of 'sinking basins', has tackled the problem differently. To avert significant imbalances he then has considered that large bodies of water may be locked up in the solid body of the Earth as constituents of metamorphic rocks (e.g. amphibolites, wherein water makes part of the molecular texture). The amount of such water held in the crust at present was estimated by Rezonor at 900 million km^3, compared with the computed volume of all the oceans at 1400 million km^3.

Summary

What is abundantly clear from postwar researches is that *all* the major oceans, and their basins, have enlarged in volume throughout Phanerozoic time. Not one of them has diminished with time. All are expanding now.

The mechanism for all this crustal expansion must be available worldwide below the lithospheric crust, and perhaps reside universally in the upper mantle.

All the mid-ocean ridges appear currently to be active, and parts of them yield geomagnetic records that cover most of geologic time since the Jurassic. There have been episodes of tectonic activity, with intermissions of quiet crust, which changes show logically in all suboceanic records — tectonic, structural, sedimentational, and denudational. These can generally be matched with events upon the continents (Chapter 10).

Whatever the mechanism of continental drift may be, it originates fundamentally beneath the lithospheric crust and is expressed *vertically* (radially within the globe). The horizontal (drift) factor at the Earth's surface is derived from the vertical as a result of spherical Earth geometry (see Fig. 31). This mechanism expresses time relations as an expanding Earth, with an enlarging graticule of parallels and meridians for measurement. This is expressed most clearly over the ocean basins. Continents do not enlarge;

they have been essentially the same size through 2000 million years. Consequently, the difference between ocean basins is apparent, and is measurable, as 'continental drift'. It is like the parade of soldiers who were moved individually (and surreptitiously) so that each was spaced from his fellows by twice the original distance. Each thought then that the others had moved away from him; but an enlargement of the space they all occupied had brought about the antipathy evident in each case.

Perhaps the best demonstration of the viscosity of the Earth is the definition of its shape under revolution. An 'oblate spheroid' is a sure indication that the bulk of the geoid deforms by flowage, probably by 'creep' in the solid state. As such, it is effective *throughout the mantle* (which is the largest part of the Earth body). The action is aided by 'de-gassing' of the mantle (Chapter 3).

And there is, of course, the puzzle which greeted the early exponents of continental drift, armed only with 'conveyor-belts' as mechanisms. 'Africa has drifted *east* from the mid-Atlantic Ridge; but at the same time it has drifted *west* from the mid-Indian Ridge. What *real* movements are involved? And what is the mechanism employed?

Fig. 31 provides the answer: for Africa, expansion of the mantle, with *vertical* displacement only. The two suboceanic ridges are likewise uplifted *vertically*; but all three are on different radii of the Earth, and the difference between any pair of them depends upon the angle between their respective radii.

Displacement of southern hemisphere continents is not north, south, east or west. It is *up*: and the general movement of the ocean floors, compared with the wandering continents, has been *down* in the oceanic basins.

But any attempt to find similar results in the northern hemisphere does not succeed so simply, for the only part of Laurasia to reach the equator is Malaya. About the North Pole itself was early an expanding region, now shown by sea floor with broken land masses (Fig. 37); but this expansion has been stopped, and locally *reversed* by northward-driving forces of global expansion derived apparently from the southern hemisphere.

On a second inspection the aggregation of northern continents encircling the pole between latitudes 60°N and 75°N with minimal opening up of sea floors expresses not only a failure to separate into diverse scattered entities, but implies the gathering together of pieces that had been scattered before.

One such encounter did occur in the late Palaeozoic, as we noted in Chapter 1 when Gondwana and Laurasia collided (Fig. 3), and when Arabia, Iran–Afghanistan, Baluchistan, and India forsook Gondwana and united with Laurasia. Each of these in turn nudged a corresponding area of that primitive supercontinent to the northward, much as icebergs do under the influence of winds and ocean currents today.

Lastly, the pear shape of the Earth, wherein the northern hemisphere appears flattened (Carey, 1976), may also be a result of that (or other) collision.

Chapter 10

Geomorphic chronology of the lands

Historical

The theory of denudational cycles interrupted by tectonic episodes of the Earth's crust, as revealed by the development in landscapes of major planations and scarps (on the regional scale both of these are independent of rock hardness and attitude), is verifiable by straightforward mapping of the cyclic denudational units (p. 183). In the early stages landsurfaces need to be related to fossiliferous strata for dating (p. 205); but this need soon passes away with experience,* and soon the landsurfaces can be recognized and classified *per se* (p. 184). Such geomorphic mapping was indeed done for small areas by a small number of geologists from the late 19th century onwards. But for full benefit, involving decipherment of prolonged topographical histories, and for proper appreciation of tectonic effects and controls, surveys covering large areas are essential.

The earliest such survey known to me was undertaken between 1928 and 1934 by F. A. Craft in southeastern Australia. Patiently, in a long series of papers, he unravelled and mapped the several cyclic landscapes of adjoining New South Wales and Victoria. I have not been able to discover that this pioneer work was ever rewarded, and to this day it does not achieve, even in Australia, the credit that it deserves.

In contrast is reproduced a modern diagram by Mescherikov (1963) showing a simple synthesis of cyclic landsurfaces in the southeast of the Russian Platform which conveys succintly quite a lot of information (Fig. 43).

In 1938 F. Dixey published a clear statement of an African sequence of four denudation cycles of Jurassic, Cretaceous, mid-Tertiary, and late Tertiary culmination respectively. Furthermore, he compared their relicts in all the

---

* For this author experience meant the acquisition of much evidence over many years, first in Africa then in Brazil and Australia and ultimately all seven continents. The results of this research were published in three books and a dozen papers. Others may now take up the study less laboriously.

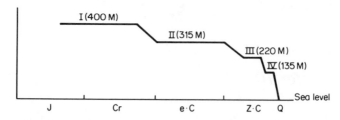

Figure 43 Classification of the planed landscapes of southeast Russia according to elevation and age. The figure expresses naturally the planation and scarp features of the landscape resulting from episodic uplifts and continuing denudation. Ages in millions of years. Compare with table (p. 187) (after J. A. Mescherikov)

territories in central and southern Africa, a sufficiently wide area to establish their reliability on the subcontinental scale. Later he extended these investigations to other African territories, Madagascar and India. The work was accepted by geologists, but without enthusiasm, They did not know what to do with it, and as presented it suffered from two defects that rendered it partly unworkable. These were:

(a) The absence of a philosophy under which multiple erosional levels with different dates of initiation could develop simultaneously in a landscape (which in those days was presumed to be undergoing universal downwearing under the ruling theory of Davisian peneplanation). This first difficulty was overcome when King (1953) postulated, in opposition to Davis, the standard denudation cycle of scarp retreat and pedimentation several of which cycles can co-exist and develop simultaneously in a continental landscape even though they have been generated successively following several distinct phases of tectonic activity and continental uplift (Fig. 43).

(b) To begin with, Dixey tied his cyclic erosional planations to definite narrow ranges of elevation above sea level. The Jurassic surface was said to stand between 7000 and 8000 feet, the Cretaceous around 6000 feet, the mid-Tertiary at 4000 feet, and the end-Tertiary at 2000 feet. Later, both Dixey and King realized independently the importance of *differential* uplift over large areas of country.

The early-Tertiary ('Moorland') planation surface of Natal rises consistently, for instance from one or two hundred metres near the coast to nearly 2000 metres on the 'little Berg' at the foot of the Drakensberg scarp (Fig. 44) (King 1976, 1982). Traced from district to district, planations have indeed been deformed like a blanket over an uneasy sleeper. But, as whatever elevation they now occur, the several planation surfaces retain their individual characteristics, by which practised observers may quickly identify them.

So a timetable involving intermittent tectonic episodes with intervening Earth planations, and covering the span from Jurassic to present day, became available (p. 187–92) for the African continent; and a map depicting their continental distribution was drawn (King, 1962/7).

In 1974, Birot and Dresch put forward a valuable comparison of planation surfaces upon the shield of Africa with those of the Atlas and the Mediterranean borders of Europe and the Levant, which have differences of tectonic style and timing (especially during the Oligocene and the Miocene in the Alpine belt). The greatest differences were found to lie between Spain and the High Atlas in the west (where strong elevations occurred at the end of the Eocene), and about Anatolia in the east.

More nearly standard were the planations of Italy and Greece (which have already been claimed as former parts of the girdle of Gondwanaland (p. 7)).

Planations around the Mediterranean are remarkable for their arid soil types, and for their wide development upon limestones.

The recognition of erosion surfaces is not difficult. The field worker in a suitable area can be introduced to the several planations as a palaeontologist learns to recognize fossils or a petrologist rocks — visually; and in a few days can continue to map the planational units with confidence (p. 184), identifying them from their diagnostic physical characteristics whatever their altitudes (King, 1976). The features of a planation surface remain recognizably the same whether it stays at low elevation or whether it is carried by tectonic movements to a high altitude. The planation may be broken by faults or tilted by warpings, but the aspect of its surface remains the same and it should be recognizable visually despite the changes of attitude and altitude, and be identifiable according to the characteristics of individual planation surfaces as mentioned in the table on page 187.

The beginner should work much upon the high ground. There he will have the elevated planation remnants about him, and there will be the best viewpoints from which accordance of summits is to be observed. The fellow who works in the valleys will have only the most recent landforms (Pliocene or later) about him. When he looks upwards he will have only a distorted view of the edges of the important high-level plateaux. Summit accordances will not suggest themselves to him.

Equally unfortunate is the man who relies, for interpretation, upon vertical air-photographs (nowadays often taken from altitudes of 9000 or 10,000 metres). He, too, has a wrong viewpoint. The smooth, level skylines of narrow mountain ridges are lost as knife edges, no matter how striking their accordance may be in the horizontal view.

Many examples have been quoted in literature of where, in a seemingly peak and ridge mountainland, the eye of an experienced topographer has discerned an accordance of summit levels signifying a former summit planation and even the manner of its deformation during uplift. Even where every vestige of the actual former planation has disappeared, the collective view over what at first appeared to be a 'sea of ridges and peaks' coordinates and, as the mind overlooks the younger valleys and gorges, there grows an

180

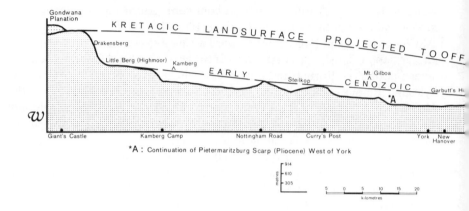

Figure 44   Denudational and sedimentational relations across the Natal Monoclin
from the Cretaceous to the Miocene. (a) In the west the highest country is th
Jurassic *Gondwana* planation (at 3350 m). The *Kretacic* landscape above the blac
Drakensberg scarp reaches 2840 m. The foothills of the little Berg carry the *Moorlan*
planation (late Cretaceous–Oligocene, at 1980 m), which formerly covered the whol
of Natal, and from which the irregular topography of the entire province has bee
derived by deep valley incisions and basins during the late Cenozoic and Quaternar
Of these, Pliocene basins are most *Widespread* from near sea level towards the coa:
to 1350 m towards the Drakensberg. (b) In the coastal crossing zone (e.g. th
Zululand coastal plain north of this section), where the denudational regime of th

appreciation of the former summit planation into which the younger valleys
and gorges have been incised.

F. E. Matthes distinguished (apparently visually) two parallel sequences
of geomorphic cycles in the Kern River–Mt Whitney and Yosemite areas of
the Rocky Mountains respectively. The first is: summit, subsummit, high
valley, and canyon cycles; the second: 'Eocene', broad valley, mountain
valley, and canyon cycles. The ages which he assigned have been partly
confirmed by K–Ar datings on some of the basalt flows poured out upon the
surfaces. These ages are: pre-Miocene, Miocene, Pliocene, and Pleistocene.
Both the sequences and the ages correspond with Planations III, IV, V, and
VI of the standard table (p. 187). Conceivably the surfaces look typical also?

Matthes may have been lucky in studying an area wherein accordances of
summit and ridge levels are more clearly expressed than usual; but his
experienced eye did not overlook their significance even in so tectonically
disturbed a region.

In the Canadian Rocky Mountains topographers have claimed to recognize
a summit planation at 900–1500 m in the west but rising eastward to 2500 m
before being eroded away in the Eastern Cordillera at elevations of 1325 m
to 1057 m. If this be a single surface, it is regarded as early Miocene ('Moorland') in age.

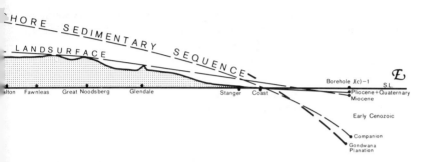

ntinent interfingers with the marine sedimentary section, early Cretaceous beds
erlie the Gondwana planation which is flexed down together with the Stormberg
saltic lavas across which it is cut. The Kretacic planation is represented by an
conformity beneath the Campanian; and the Moorland surface is beneath a shell-
d of lower Miocene (Burdigalian) age. The Widespread surface is represented by
e coastal plain, formed by marine transgression to 180 m in places. (c) 24 km
fshore from Stanger beach, borehole J(c)-1 (Fig. 49) recorded beneath an upper
igocene–lower Miocene shelly bed more than 1312 m of early Cenozoic sediments
ove a great wedge of upper Cretaceous, all of which afford clear correlations with
the denudational features of inland Natal

Younger, lower surfaces, are all regarded as Neogene, we suggest late
Miocene, Pliocene, and Pleistocene in age respectively.

Applied to planation surfaces the phrase 'visual recognition' raises two
special queries:

(a) Is it possible to distinguish visually one planed landscape from another
of different age and significance.
(b) Does there exist a geomorphic historical time scale that can usefully be
applied to planation landsurfaces?

The answer to each question is a qualified 'yes'.

The chief qualification arises from the heterogeneous nature of the Earth
whereby even forces of truly global application may be (and are) resolved
with local differences. The behaviour of orogenic belts, for instance, differs
from the response of cratonic areas; and minor differences are therefore to
be expected from place to place in:

(a) the operation of tectonic forces both in style and time; and
(b) the denudational expression of quiet intermissions between tectonic
episodes, including available relief, reflection of bedrock factors, and
facies subdivisions influenced by prevailing climates.

In Japan, as elsewhere around the Pacific Ocean, tectonic and volcanic movements have been almost continuous throughout the Cenozoic era with many local phases of activity.

In northeast Japan summit surfaces range between 400 and 1000 m in altitude with significant planations between 200 and 400 m on the Kita Kami Mountains. Younger surfaces are widespread between 80 and 18 m in Honshu. Offshore the Japanese Trench provides a record of negative Pleistocene movements.

In southwest Japan the oldest surfaces (pre mid-Miocene) are much warped, and the three lowest Pleistocene terraces are matched by submarine planes at 100, 900, and 1800 m depth. Long and patient study is needed in such a region.

The age of the 'Moorland' planation surface (early Cenozoic), for instance, is given by definition (p. 188) as from late Cretaceous (when the previous landscape was elevated and the denudation leading to the Moorland planation was initiated) until the beginning of the Miocene when the lands were uplifted once more and the denudation of the succeeding 'Rolling' surface was begun. Because of the long interval of time and the vast area over which the Moorland cycle was current it admits of great local differences of time at which the cycle began and ended.*

Even after a new cycle of landscape development has begun near the shore, an interval of time is required before it spreads over the continental interior and destroys the features of its predecessor. So at the present time the Moorland planation, the development of which was succeeded in the coastlands by the Rolling surface during the early Miocene, may sometimes still be found evolving upon the watersheds of an inland terrain. The Rolling surface then perhaps appears upon shoulders and benches of the upland spurs while lower downstream are the Pliocene basin plains and Pleistocene gorges of the yet younger Widespread and Youthful denudations (p. 191). All four cycles thus operate concurrently, but in strict order of precedence in different parts of the total landscape, and each exhibits its normal features by which it can be visually distinguished. A help in this regard is if the intervening episodes of uplift afford a large measure of available relief, so that scarp development and retreat separate the several levels of planation distinctly.

In the interior of Australia, which has only a small measure of continental relief and correspondingly few distinctive cyclic scarps, Mabbutt gives the age of the Great Australian Planation as late Cretaceous to Pliocene. But in the eastern belt of tectonic relief the several cyclic surfaces are more distinct from each other, and were long ago correctly identified and mapped by Craft.

In a few specially favourable localities, e.g. Drakensberg, Nilgiri Hills of Southern India, and the states of Rio de Janeiro, Minas Gerais, and neigh-

---

* A similar concept bearing upon the time taken for the distribution of fossil species is *homotaxis*.

bouring Espiritu Santo in Brazil, several of the six major continental planations rendezvous and can be identified within the compass of a few score
kilometres.

This is no more than the usual stricture of local variation that applies to
all geology. Formations, fossil faunas, states of igneous and tectonic activity
all differ naturally from place to place; but geologists succeed nevertheless
in making correlations and stating broad generalizations for any stage of
geological time.

While in the history of geology knowledge has often proceeded from the
local to the general, later generations of students are taught the generalizations and interpret them locally.

The ages of planed landsurfaces

The dating of planed landsurfaces in terms of the geological timetable must
now be defined. Each such landscape begins following an episode of tectonic
uplift. Thereafter it may continue to develop locally for an indefinite period.

But eventually another tectonic episode generates a newer cycle of erosion
which in turn develops by scarp retreat and pedimentation — at the expense
of its predecessor.

In suitable regions a number of cycles, begun at different times, each
consuming the landscape of its predecessor and in turn being consumed by
its successor, march in order across the continents from the coast towards
the interior.

As each intermission of denudation is initiated by a tectonic event which
can normally be dated by ensuing sedimentary and palaeontologic data at
the coast (especially upon coastal plains) the way is clear to correlate events
upon the lands with events beneath the sea. As each denudational landscape
is consumed it provides land waste which builds the sedimentary record of
its successor.

The geological timetable is constructed from the same data, so the dating
of planed landscapes is merely another facet of the same chronology. The
tectonic episodes appear as unconformities in the coastal sedimentary
sequence.

More rarely, continental basins contain sediments that enable surfaces to
be dated locally.

So in 1948, when the number and distributions and dates of planation
cycles in Africa had been ascertained by field survey and a table of events
had been established that was applicable to the entire continent, the present
writer posed himself the following global problem: do the several continents
each display recognizable suites of planational landscapes, and does each
continent have its own individual suite recording an individual tectonic and
geomorphic history? Or is there a single recognizable overall plan of
tectonism and ensuing denudations similar in all the continents? Put another
way: the several continents are thousands of miles from each other and could

thereby have individual histories: but all the continents are on one Earth and perhaps the basic tectonics which initiate and end the planational intermissions are under global control.

After prolonged library search for relevant data, Brazil was first examined, in 1954. The start was dramatic — in the first day's fieldwork from Belo Horizonte four out of the five planations then known in Africa were identified. In the ten weeks that followed (and after 22,000 km of travel in a *Jeepi*) an area had been mapped in planation cycle units which was three times the size of Britain with its outlying islands. This mapping showed, moreover, that the surfaces had been tilted and warped in large amount by vertical tectonic movements exactly like those of Africa. Two rift valleys on the crests of great arches were identified. In one flows the Rio São Francisco, in the other the Rio Paraiba (Fig 12).

Brazil and Africa, even though separated by the width of the Atlantic Ocean, have operated in concert by vertical tectonics and have experienced the same interludes of cyclic denudation throughout the long period spanning from the Jurassic to the present day (see King, 1962/7).

Be it noted: this remarkable parallelism of geomorphic history between two separate continents now far apart was not like the close comparisons between many distinctive rock formations that had been used by du Toit to demonstrate that the formations had accumulated when the two continental masses were formerly contiguous. The respective multicyclic landscapes of Africa and South America developed only *after* the continents had separated! Their similar geomorphologies had evolved by denudation in response to similar vertical displacements in the two continents *subsequently to the break-up of Gondwanaland and continues to the present day*. A return to Fig. 31 solves the problem. The continents of Africa and South America are today widely separated; but on an expanding globe they both operate under similar *vertical* controls! They furnish a final proof of the validity of an expanding Earth, and demonstrate, once again, by what mechanism horizontal continental drift is achieved!! My problem of 1 January 1948 was solved in a way that I had not expected!!!

Further surveys in different parts of the world yielded always the same outline of landscape evolution, though often providing minor or local variations according to the regional lithology and climate, e.g. Antarctica (King, 1966).

All continents, therefore, exhibit the same chronology of planations with semisynchronous tectonic interruptions between them (King, 1962/7, 1976). This is proof that: (a) the displacements are, in every case, vertical; and (b) the tectonics are global. The amount of displacement naturally differs from place to place, but all the lands are normally uplifted. Of course, basins do form (e.g. Congo, Lake Eyre); but these are usually relative, and form by greater uplift of the basin rims rather than by depression of its centre (e.g. Kalahari, Chad).

In the interludes between tectonic uplifts, denudation operates to produce, by scarp retreat and pediplanation, newer planations at lower levels. The late Mesozoic to Recent history of the widely sundered southern continents is clearly displayed as a succession of vertical epeirogenic and cymatogenic uplifts. Such a history must find logical place in any acceptable hypothesis of global tectonics (Melhorn and Edgar, 1976).

After fourteen years research in many lands, the results were published in *The Morphology of the Earth* (1962/7). The second premise (posed in 1948) is true: *despite the admitted complications introduced by local factors, a single simple global pattern of tectonic episodes with planational intermissions is recognizable for all the continents from the mid-Mesozoic until the present day*. Naturally the history is best displayed in cratonic areas of simple, widespread uplift, least in the deformed orogenic belts wherein only the latest chapters in the history are likely to be encountered. Of all the continents, Africa is the one in which the several planational surfaces may best be seen and mapped, and their history determined (King, 1962/7).

After ten more years of research ranging through all seven continents, from the Arctic Circle to the South Pole, the writer was ready to select the world standard section, which he did in *The Natal Monocline: explaining the origin and scenery of Natal, South Africa* (1972/82). In his home province, in the section from the Drakensberg to the sea (Fig. 44) was displayed a complete history of landscape development from Gondwana time (before Africa took status as an individual continent) to the present day (Fig. 2). All this was beautifully displayed in tilted planational steps upon a monocline 250 km wide which had repeatedly tilted to the southeast during Cenozoic time about a hingeline situated offshore of the present coast. At the Drakensberg in the west, cumulative uplift amounted to thousands of metres so that the several planational landscapes are there separated from each other by large, scarped intervals (Fig. 45(a)).

Moreover, although the planation surfaces are lost to view beneath the Indian Ocean beyond the Natal coast, to the north the sea floor has been raised in the Zululand coastal plain. In this plain are laid out the several marine sedimentary formations corresponding to the planations of the land to the west, ranging from the earliest Cretaceous to the present day. As each of these formations is fossiliferous they give the ages of the several planations in turn (Chapter 11). Denudational and sedimentational records are adequate for all necessary correlations (King, 1972) and I know of no better section to provide a standard for the geomorphic development of the world's landscapes from the mid-Mesozoic to the present day. The land area from which it is derived (table 0) has been continuously land from Jurassic time to the present day. Herein we do not deal with planations which may have been buried for an interval of time and later have been resurrected under new phases of denudation.

The several global geomorphic planations are now tabulated (Table 1), described and dated in order of their occurrence (King, 1976).

186

Figure 45(a)   The face of the Drakensberg, between Giant's Castle and Cathkin Peak, showing planations:
(a) the Jurassic Gondwana planation above the snowy scarp on left (3350 m);
(b) the Kretacic landscape above the black Drakensberg scarp (2840 m);
(c) on the bevelled foothill spurs the Early Cenozoic Moorland planation (1980 m);
(d) on the distal ends of the spurs the Miocene Rolling Surface (1830 m);
(e) Pliocene valley forms (1350 m above sea level)
(Survey Dept, University of Natal)

Figure 45(b) The coastal hinterland of Natal. Amid deep dissection by Pleistocene valley forms (Valley of a Thousand Hills) are numerous summit remnants of the Moorland planation. These rise westwards (away from the observer towards the 'little Berg' at the foot of the Drakensberg. The erosional planation cuts impartially across all the geological formations of Natal (Survey Dept, University of Natal)

The several global planations cycles and their recognition

	Planation cycles	Recognition
I	The 'Gondwana' planation	Of Jurassic Age, only rarely preserved
II	The 'Kretacic' planation	Early-mid Cretaceous Age, on certain very high plateaux and ridges in Lesotho
III	The 'Moorland' planation	Current from late Cretaceous till the mid-Cenozoic. Planed uplands, treeless and with poor soils. Often the oldest planation identifiable in Natal. Occurs *below* the Drakensberg escarpment. At the coast is overlain by marine Miocene strata
IV	The 'Rolling' landsurface	Mostly of Miocene Age, forms undulating country above younger incised valleys
V	The 'Widespread' landscape	The most widespread global cycle, but more often in basins, lowlands and coastal plains than uplifted by recent tectonics to form mountain tops. Pliocene in age
VI	The 'Youngest' cycle	Modern (Quaternary in age) represented by the deep valleys and gorges of the main rivers. Sometimes has glacial features

I The 'Gondwana' planation, of Jurassic Age

The history of geomorphic planations began before the continents assumed their present outlines and positions; when instead they were integral parts of two parent supercontinents — Gondwanaland and Laurasia. Those super-continents were themselves widely planed, e.g. across Drakensberg basalts, and when they broke into continental fragments each of the daughter continents began its own history with a widespread 'Gondwana' or 'Laurasian' landscape as its surface expression. In the cratonic areas of South and East Africa, southern India, Brazil, and elsewhere small relicts of this planation still survive upon the highest and most ancient watersheds. There, high above the well-planed 'Moorland' early Cenozoic planation (III), they have been continuously exposed to the winds and the weather since they first took shape in Gondwana at least 170 million years ago. Naturally, such relicts are very rare in existing landscapes; yet despite their great age, *they have not been buried and resurrected at any time in their history*.

Towards the coast where these surfaces have been flexed down monoclinally, this planation (still on Drakensberg basalts) passes below late Jurassic or early Cretaceous strata (both continental and marine) which serve to date the Gondwana surface as Jurassic. At a few places the original Gondwana landscape consisted of desert sandstones, and these contain dinosaurs also of Jurassic type, e.g. *Massospondylus* at Gokwe in Rhodesia (Bond and Bromley, 1970; Lister, 1976).

Incidentally, middle Jurassic marks the beginning of modern sea-floor development, and the beginning of modern land geomorphologic history. It was the time of continental break-up in both Laurasia and Gondwanaland, and saw the separation of the modern continents start the story of continental drift. Conversely it saw the obliteration of much evidence relating to earlier regimes both by land and sea. Globally it was like a new creation.

Active episode A

The tearing apart of Gondwanaland and Laurasia not only created the modern continents, it also provided new base-levels for each continent.

II The 'Kretacic' (or 'post-Gondwana') landscape, of early and mid-Cretaceous Age

New planations for individual continents began dissecting and destroying the inherited 'Gondwana' and 'Laurasian' landforms. These 'Kretacic' planations, of Mesozoic age, are also rare nowadays, and they also appear upon elevated plateau areas and upon ancient watersheds where they may have a relief (with steep scarps) of several hundred metres. Soils on basalts are commonly red due to prolonged oxidation of the iron minerals.

The resulting debris shed from the lands accumulated around the continental borders as late Jurassic to mid-Cretaceous marine sediments. In some places this landsurface is overlapped by Cenomanian deposits.

Active episode B

Vertical uplift, effective over whole continents, prevailed during mid-Cretaceous time, introducing new base levels almost everywhere. Conversely, local basins were formed, e.g. in Western Europe. These are marked by Cenomanian transgression. This was succeeded by:

III The 'Moorland' planation (late Cretaceous to early Miocene)

For this exceedingly long interval of time, tectonic quiet reigned over vast areas *outside the belt of Alpine orogenesis*. A planation of extraordinary smoothness developed over enormous areas in *all* the continents. In many places it was encrusted with a senile soil profile of laterite, calcrete or bauxite, which is normally diagnostic, although younger and less complete versions of these soil types are known also upon the succeeding planation.

In their variety these aged residual soils reflect the chemistry of the bedrock upon which they lie, but they all belong to the same planation cycle, whatever its post-Oligocene history has been. The enormous bauxite deposits of Weipa, Australia, for instance, crop out over the early Cenozoic planation which, stretching far and wide, there sinks down to pass below the sea in the Gulf of Carpentaria. Quite different topography, however, marks the interior of Surinam, where the bauxite deposits occur upon planed watersheds where the Moorland surface was uplifted during the late Cenozoic and has been dissected by deeply incised, younger valleys. Despite the contrasted topographies of the two regions, which at first sight appear to have little in common, the geology of the bauxite deposits is the same in both.

Absence of these duricrusts elsewhere does not disprove 'Moorland' planation; they have either not been present originally, or have been removed under later superficial erosion. Many areas of Moorland planation bear no signs of ever having possessed duricrust and may be deemed to have been climatically unsuited or to have lacked the necessary constituents in the bedrock. In calcareous terrains, extensive cavern systems developed below the water table, for example, in the Transvaal.

Outside the rare areas in which the older Gondwana and Kretacic planations (I and II) can be identified, this 'Moorland' planation is to be sought upon the highest, and often bleak, plateaux. It has also been given many local names, for example the 'African' planation, the 'great Australian denudation cycle', the Schooley peneplain, the English 'Moors', and so forth. With emphasis upon its extreme planation, it appears particularly flat when viewed in cross-section as across the intervening valleys of succeeding cycles which commonly intersect it. From it, most of the world's present scenery has subsequently been carved by renewed erosions.

To ascend the steep slopes from younger, incised valleys and emerge not upon a mountain peak but on a flat plateau or 'moor' of some earlier planation with its horizon like the sea and nothing but the sky above, is always impressive to a geomorphologist. Well may he meditate upon its ancient significance.

In some places where it has been affected by (usually minor) tectonics the surface has been broken up or has developed in subphases.

Corresponding sedimentary formations begin with abundant late Cretaceous, diminishing volumes of Eocene, and often cease before the Oligocene as the lands were planed (supplying little land waste, especially after duricrusting).

This cycle of planation terminated at the onset of Alpine orogeny during the late Oligocene.

Active episode C

Is represented in the mobile belts by the main Alpine orogeny. In the cratonic areas it is marked by uplift and tilting of the earlier landscapes.

IV The 'Rolling' surface, Miocene in age

So new continental planations developed widely to Miocene base levels. Much of the smooth 'Moorland' planation (III) was destroyed and at lower levels a new 'Rolling' landscape (IV) succeeded. Relicts of the earlier 'Moorland' planation not infrequently stand at strong scarps abruptly above the

'Rolling' landscape; in fact the triad of two planations (the upper much the smoother) separated by clear scarps can be visually identified and should be sought in continental landscapes. Equally typical is where the smooth older planation is dissected by shallow, headwater valleys of planation IV which may compose most of the landscape (for example, the highveld of South Africa). Patches of the typical old age residual soils upon the very flat watersheds serve to identify the Moorland planation with certainty (for example, the 'clay with flints' of southern England, and the older laterite of India). Subsequent deformation of the 'Rolling' surface usually follows the pattern set by that of the older 'Moorland' surface, so that the two are often found in close proximity.

The sedimentary deposits associated with the 'Rolling' planation begin with the early or early–mid-Miocene.

Active episode D

About the end of the Miocene period the first of two mighty upheavals affected the continents. The upwarpings are generally linear; those areas which lagged in elevation generally make very broad depressions. Uplifts along the axes were of the order of hundreds of metres.

Many of the earlier drainage patterns were disturbed and some were entirely altered, even along the prior positions of continental divides. The previous planations (III and IV) were often carried to high, and even mountain, levels where they came under very active denudation.

V The 'Widespread' planations of Pliocene time

Following active episode D, and during the Pliocene period, the landscape that developed by scarp retreat and pedimentation is the most widespread of all the cyclic planations. Characteristically this landscape is expressed in the form of basins closely related to modern drainage lines; but there is no simple relation between the various basins, and independent basins at two or three different elevations may often be developed even within a single, major river system, especially where much tilting was inherited from active episode D. Two subphases can usually be distinguished; and the pediplains of this cycle are noticeably more horizontal than those of the older (and more tilted) planations.

In the mountain belts of greater uplift, Pliocene valley development considerably reduces the area of formerly continuous planations (III and IV) to isolated remnants at high elevations, and where uplifts were late this Pliocene surface itself composes the summit planation, e.g. Andes, Southern Alps of New Zealand.

About many coasts subsidence occurred and offshore there developed a wide marine-cut Pliocene shelf, the deposits on which are fossiliferous sands and calcarenites. Before the end of the Pliocene period the sea retreated leaving a broad coastal plain. In the continental interiors, depressions (or areas which failed to rise tectonically with the watersheds) became foci of local deposition (basin plains). Three examples from Africa are: the Kalahari, Congo (Zaire), and Chad.

Five characteristics usually serve to identify these landscapes:

(a) the scarps are generally in a state of active development and the corresponding pediments are the most widespread of such features in any landscape;
(b) the planations are the most nearly horizontal because they have been least tilted or warped by earth movement;
(c) the planations are only partial within the landscape of broad regions. They are relatively low in elevation and related to existing river systems;
(d) if younger landforms be present, these will be Quaternary incisions and terraces; and
(e) two phases of development and occasionally three occur within the compass of local (Pliocene) development.

In total, Pliocene landscapes cover a greater area than any other. This is true upon each of the continents with the possible exception of ice-covered Antarctica.

Active episode E

At the end of the Pliocene period a second mighty upheaval enhanced the differential movements of episode D and also produced strong seaward tilting of continental margins. The mountains attained their present altitudes.

VI The 'Youngest' landscapes, of Quaternary time

Here belong all the late Pliocene to Recent valley incisions, terrace cuttings, glacial phenomena, with modern phases of scarp retreat and pedimentation that provide much of the detail in existing landscapes. Too recent to have produced extensive planation, this intermission is, on the contrary, responsible for much of the ruggedness that characterizes the modern aspect of the Earth (Fig. 45(b)) in contrast to the more extensive planations typical of earlier geological ages.

Denudation during this regime has instead greatly reduced the remaining areas of former planations, especially where they occur upon high mountains created by active episodes D and E.

Clearly, geomorphologists who concentrate their attention upon rivers, valleys, and lowlands will normally encounter landforms of the later planation cycles. Only those who range far and wide and climb to greater altitudes are likely to become familiar with the older, ancestral planations. A subcontinent is a good range, and 'from the mountains to the sea' affords a good section for the study of geomorphic chronology.

Understandably, this table of planations agrees very well with the long-established stratigraphical timetable for Cenozoic time. Both are based on conditions and developments following the same global tectonic upsets! In my experience (King, 1962/7, 1972, 1976) (and see Chapter 11) it has operated in all the continents of the world, with due allowance for minor differences from precise synchroneity and amounts of local uplift and denudation which are in accord with crustal inhomogeneity. The controlling subcrustal forces appear to act similarly on a worldwide scale with emphasis upon *vertical* displacements.

Mapping of the world's major planations on the basis of the given table (p. 187) is a distinct and useful possibility. The study should always be on the scale of broad regions and not cluttered with details of minor or local landforms. A convenient scale for field mapping is 1:500,000, with publication at 1:2 million or 1:5 million, although this sinks much of the detail of older planations.

One difficulty for continental (or world-scale) compilation is the present lack of trained observers who *think* on a continental scale.

The significance of this geomorphic chronology is that it demonstrates, with the utmost clarity, that geotectonic events to which geomorphic evolution is due, are not continuous but episodic in time, short episodes of activity alternating with much longer intermissions of tectonic quiescence during which denudation and deposition are dominant. The record may be read both from denudational landforms and/or from the stratigraphy of the oceanic (and occasionally terrestrial) deposits.

In practice, the denudational record is found to be more 'stepped' than the sedimentational one in which 'clumps' of disconformities indicate perhaps five times the number of interruptions by minor movements. The lands, of course, record these too, but the processes of denudation are such that larger phases tend to 'overrun' smaller earlier ones and to incorporate them into a single erosional pulse of greater cyclic significance.

The mechanism of vertical uplift upon diverging radii of an expanding spherical Earth not only explains the effect of continental drift by which the continents scattered but also explains the continuing similarities of geomorphic development from continent to continent as time, and drift, proceed. Indeed, with such a mechanism geomorphic analogies may be expected to continue into the future.

Seeking cyclic surfaces

In any given geomorphic survey, seek first to find, either upon the tops or the flanks of the highest country, traces of the very smooth 'Moorland' planation. This is the easiest planation to identify by eye. Its aspect is usually treeless and barren, its soils are old, and it is commonly diversified by shallow valley-heads of the succeeding 'Rolling' surface. Familiar examples are: Dartmoor and Exmoor, Ilkley and the Yorkshire Moors, the Baraques of the Ardenne, the highest part of the Central Plateau of France, and the Schooley surface of eastern North America. In the southern hemisphere examples are: the highveld of South Africa, the Great 'Australian' planation with its laterites and bauxites, and the 'sul-Americana' landsurface of Brazil.

Although very smooth, especially in profile, none of these is horizontal; all of them show a distinct slope (either seaward or to inland basins) of about one degree which has been induced by late Cenozoic tilted uplift. Even where the actual planation has been largely destroyed by later erosion (as in a small country like Britain) its former presence may sometimes be demonstrated by plotting the maximum elevations within the squares of a grid and drawing contour lines thereon (Figs 46, 47).

In limestone country cavern systems were commonly excavated beneath the water table of the 'Moorland' planation. As the area was drained by the incision of valleys during later denudation cycles, the caverns were opened up, and partly dried out. A curious result of this process is that older sediments are sometimes lowered in sinkholes beneath the level of younger planations, e.g. late Cretaceous lignites in sinks beneath the early Cenozoic, Schooley surface of the Appalachians (Pierce, 1965).

If no trace of the Moorland planation can be found, only younger and more irregular landscapes are presumably present. Of these, the 'Rolling' surface often makes the high country above broad basin plains typical of the 'Widespread' (Pliocene) cycle. This combination is found over much of Germany and France; and is a common landscape in all the continents.

Occasionally, in orogenic zones, the younger denudations have been uplifted to high altitudes (e.g. Andes and New Zealand Alps) such overelevated Pliocene features are flanked by deeply incised systems of Pleistocene gorges (see *The Morphology of the Earth*).

By experience, seeing, and elucidating many landscapes and extrapolating from one to another, the observer not only acquires facility in recognizing the characteristics of the several planations, but also becomes familiar with the nuances and variations which they display in different regions and environments. The Earth is heterogeneous, physically, and there are many variables which cause departures from a rigid style of development and landscape evolution. Discipline is exercised by the strict control of mapping the several planational units.

But mapping is not ideal. It is often difficult to maintain a scale. So much of the landscape may belong to the later 'Widespread' landscape cycle that remaining, distinct remnants of older landscapes are reduced to pinpoint size upon the map. I well remember the northern end of the Rhodesian plateau ending at a mighty scarp leading down to the lower country towards the Zambezi. Fifty miles away, across the flats, rises the conical hill of Showe, its *top is bevelled by a 2 acre remnant of the main plateau*.

Cenozoic planation makes the smooth watersheds of the South African highveld, *with westward dip*. So it descends beneath the sands of the interior Kalahari basin. Beneath these Pleistocene sands the early Cenozoic (Moor-

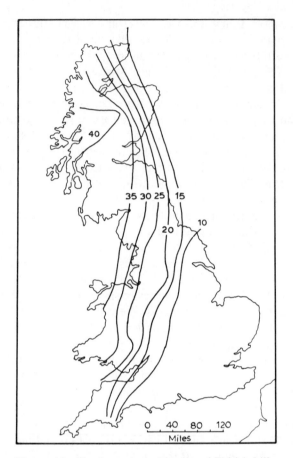

Figure 46 'Farthest east' altitudes of British hills
and mountains demonstrating: (a) the former existence of a summit topography of small relief (even
though scarcely any of this topography remains
today; and (b) that this former topography has
hinged along a north–south axis passing through
East Anglia, giving progressively more uplifted
country towards the west (Wales) and a depressed
area (North Sea Basin) to the east. Elevations in
hundreds of feet (after D. L. Linton)

land) landscape is found as the aggradational surface of an older series of
depositions (the Kalahari Marls) which are correspondingly of early Cenozoic
age topped off with calcrete duricrust. In southwest Africa the planation
emerges from beneath the desert sands once more and *rises* westward to
1500 metres at the foot of the Naukluft Mountains. And beyond the mountains, across the Namib it descends yet again as a coastal monocline to pass
offshore beneath the sea (Fig. 7). At both east and west coastal plains this
planation passes beneath lower Miocene marine fossiliferous deposits. So

Figure 47   Primitive drainage system of Britain.
Compare with Fig. 46 (after D. L. Linton)

between them, its altitude affords a record of vertical, late Cenozoic, differential uplifts.

Many examples of such tectonic displacements were adduced in *The
Morphology of the Earth* (King, 1962/7). With such information from the
lands needs to be assessed related information from the sea floors (preserved
there in the sediment piles now being tested by deep-sea borings). A critical
transitional zone here is the continental margins (see Chapter 11).

A review of modern published work on the manner in which planation
develops a suite of landsurfaces, either sequentially or simultaneously, has
been offered by Dr Lidmar-Bergstrom (1982). The principal processes are:
(a) etching, (b) stripping of saprolite, (c) pedimentation, and (d) duricrusting.
The proportions of each are said to be governed by tectonic upheaval,
lowering of base levels, or climatic change. To which I would add — from
many years study of bornhardts — type of bedrock. These have been found
under all climatic regions from tropical to polar.

There has been much discussion of deep weathering in granitoid rocks.
My own experience which includes twenty miles of tunnelling through such
rocks in the Valley of a Thousand Hills, is quite to the contrary. Deep

weathering seldom goes beyond 10 m in depth. Inspection of aqueducts after 15 years for roof falls revealed only two such falls, of about half a barrowload each, and these could be attributed to failure of the contractor to bar off every vestige of loose rock after blasting.

In Brazil schists on the other hand, showed advanced deterioration continued beyond 50 m depth.

Scarp retreat

The scarps that separate the different planations in the landscape provide examples of a phenomenon not considered in the Davis theory of peneplanation. Under denudation, scarps do not flatten indefinitely, they assume declivities which are decided by the nature of the rocks forming the scarps, and the manner in which the agencies of erosion act upon the scarp face. Generally, scarps modelled principally under the influence of water flow become seamed and gullied and retreat under the attack of gully heads and the widening of ravines. Other parts of scarps that are composed of weaker materials are generally subject to landslips and other mass movements. The scarp face of the Drakensberg, which has retreated by more than 150 km during the Cenozoic era, preserves an amazing steepness in its basalt face, and the Cave Sandstone below the 'little Berg' remains impregnable to climbers for much of its outcrop. But the underlying Red Beds and Beaufort shales present mainly slipped topography on the lower slopes. The 'critical height' for each material thus governs the aspect of its outcrops.

The Mesozoic 'Gondwana' and 'Kretacic' planations, above the black precipices of the Drakensberg, show no significant signs of general lowering during Cenozoic time (Fig. 45). They are, verily, separated in level by steep scarps which have plainly retreated from the sites of stream incisions (Mesozoic). But the two planations are themselves still *flat*, and what agency is there to modify a surface which is already flat? The Davisian hypothesis of peneplanation is here, once more, shown to be in error — by topography on the grandest scale.

The Drakensberg is almost unique in that its bedrock structure is flat and largely original. So, under erosion it makes a wall, not peaks (Fig. 45(a)). The ascent (say of Sani Pass) takes one from the modern valley (Pliocene cycle) past the Cenozoic-bevelled spurs of the 'little Berg' (bearing the 'moorland' planation) to the sparsely inhabited summit and subsummit Mesozoic landscapes, bare and brown/red, of high Lesotho (Fig. 7). A sensitive observer of landscape may be forgiven for feeling at the high levels not only that he makes a passage in altitude and in climate, but also one in time — that he is back in the Mesozoic! He would be more surprised to see African elephants than to see dinosaurs.

Chapter 11

A key to Earthscapes: continental margins

The tectonics of continental margins

At continental margins the surface of the lithosphere disappears beneath the ocean, where nowadays the main features of its submarine form are becoming known, with details which permit correlation with the known configuration of the lands (Fig. 48). The study of Earthscapes has begun.

Many authors have commented that the history of the present sea floor begins in the Jurassic with local basin formation and the accumulation of sediments (sometimes fossiliferous) of the same, or early Cretaceous, age.

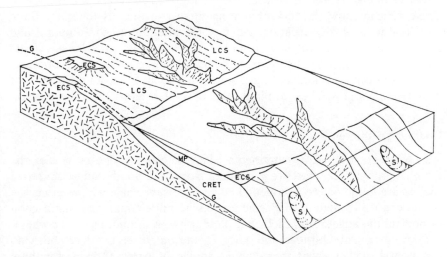

Figure 48 Geomorphic homology between coastal hinterland and continental shelf. G, Gondwana planation; ECS, early Cenozoic (Moorland) planation; LCS, late Cenozoic (mostly Pliocene) landforms and unconformities; Pleistocene gorges are continued as submarine canyons of the same age. For clarity the Kretacic and Rolling (Miocene) landscapes are omitted, although examples of both are well known. S, slumping on the continental slope

None that I know of has remarked that the denudational and sedimentational history of the current lands begins about the same time; it does. The
oldest surviving denudational planations are Gondwana (mid- to late Jurassic)
and contemporary terrestrial formations were accumulated in continental
basins of the same age (e.g. the Gokwe Beds of Rhodesia, with *Massospondylus* (Lister, 1976)); and many other contemporary sets of red beds are
known in other parts of the world.

Over both continents and sea floor alike all previous records of gemorphic
development at the surface of the lithosphere were either destroyed by
erosion or buried stratigraphically about the mid-Mesozoic. This was also
the climax of collision between Gondwana and Laurasia, resulting in those
continental disruptions and travels that were typical of the later Mesozoic,
and were in turn succeeded by the general placidity of early Cenozoic time.

In both subaerial and submarine environments, geomorphic evolution has
since proceeded in cycles (either denudational or sedimentational), and each
cycle has followed an episode of tectonic unrest which created new topographic states. The continental side of the picture was worked out between 1930
and 1960 (Chapter 10); but only recently have sufficient data become available to start filling in the detail of sea-floor history. Now the first steps can
be made toward devising a unitary evolution for the lithospheric Earthscape
during the past 170 Ma. In this, naturally, studies in the region of the
continental–oceanic margins assume great importance.

In Chapters 9 and 10 we reviewed: (a) records of stratigraphic sedimentation in the ocean basins; and (b) denudation chronology upon the lands.
There remains now the task of putting these two records together. Basic
correlations need identification and tectonic interpretation. Temporal and
tectonic identities and correlations in the zone of coastal hinterlands, coastal
plains, and continental shelves need to be established. Already, important
correlations can be made, and great similarities found, over long distances of
continental margins. These are indicative of similar (global) tectonic control.

Continental shelves

All coasts are fringed by continental shelves. Narrow shelves border the
more recently folded coasts (e.g. western North and South America), broad
shelves are found where continental edges have sunk shallowly, as along the
Arctic coast of Siberia. On Pacific aspects island festoons may stand upon
or beyond the edge of the shelf, but these festoons are not usual elsewhere.

Prior to the late Jurassic and early Cretaceous there can have been no
continental shelves about the Atlantic, Indian or Arctic Oceans, for these
oceans did not themselves exist. About the Pacific Ocean powerful Cenozoic
diastrophism interfered with any pre-existing continental shelf, and it is
doubtful on available evidence whether the modern Pacific shelves can boast
an ancestry much beyond the same Cretaceous Period. Nevertheless, when
New Zealand was a geosyncline east of Australia and the Tasman Sea did

not exist, i.e. in late Triassic time, there accumulated beds of boulders and gravel *from the west* that may now be inspected at Marokopa on the Waikato coast. The rock types, including granite, must have come from Australia, or the Lord Howe Ridge, or from a belt of early Palaeozoic rocks formerly extending northward from the South Island.

Recoveries from cores drilled into the continental shelf include representatives of all the major geologic epochs from earliest Cretaceous to Recent. Though the pattern varies from place to place, Cretaceous rocks seem consistently to be recovered from the outer and lower parts of the shelf, with Cretaceous, Eocene or Miocene formations at the break of slope to the ocean floor. Younger formations lie most thickly upon the inner and mid-sections of most shelves (Fig. 48), but they do occur as gravels and sands upon the edges of both the North American and European shelves, also upon the extremity of the Agulhas Bank, South Africa, and along the eastern coast of Australia where, according to Fairbridge, 'great slabs of beach rock have been dredged up from 7–75 fathoms'. Shallow water fauna in these slabs suggest a Plio-Pleistocene age.

Where the basement occurs beneath a cover of early Cretaceous formations we may recognize the surface of unconformity as equivalent to the Gondwana or Laurasian planation upon the lands, and the succeeding formations, and the unconformities between them, continue to record denudational and tectonic effects within adjacent coastal hinterlands until the present day (p. 201).

In this way coastal hinterlands and shelf areas show remarkable geomorphic homologies throughout a long history.

There arises, however, the question as to what marine agency was responsible for the levelling of the shelf in early Cenozoic time, a levelling that was preserved, with minor modification, until the offshore canyon cutting of Quaternary time? Briefly the shelf is too wide, and towards the outer edge too deep, to have been controlled by normal wind-generated waves of the ocean surface. There are few reliable data concerning emergence of the shelf caused by the swinging sea level of the Ice Age! But even allowing for quite unproven large eustatic withdrawals of the ocean, the period was too short to explain the essential levelling of the shelf, though it may account for many known minor features.

For explanation we turn to researches by Petterson, finely summarized by Carson (1953): 'As the ocean water presses in toward (the Baltic Sea) it dips down and lets the fresh surface water roll out above it, and at that deep level where salt and fresh water come into contact there is a sharp layer of discontinuity, like the surface film between water and air. Each day Petterson's instruments revealed a strong pulsing movement of that deep layer — the pressing inward of great submarine waves, of moving mountains of water. The movement was strongest every twelfth hour of the day, and between the twelve-hour intervals it subsided. Petterson soon established a link between these submarine waves and the daily tides. . . . Some of the deep waves of the Gulmarfiord were giants nearly 100 feet high'.

Similar internal tide waves, strengthening toward the lands, have been recorded at 40, 160, and 320 miles off the coast of California, and J. L. Reid considered that the internal wave originated with the tide near the coast.

Surely, in the great internal tide waves that surge twice daily from the ocean depths over the shallow shelf areas under the influence of lunar instead of terrestrial gravity lies the agency responsible for levelling the continental margins and their development, during the long period of early Cenozoic quiescence, into the basic form of the continental shelf.

Inspection of sedimentary records from the shelf then show that the dip of strata, and the slope of unconformities, are greater than might be expected from the action of oceanic agencies alone. The formations and unconformities have been tilted seaward (monoclinally) at intervals during the later Cenozoic. There have been repeated tectonic episodes: always in the same sense — the lands go up, the sea floor down (Fig. 49).

The theory of a long interval during the early Cenozoic without drift of continents, sea-floor spreading or mantle convection is supported by much sedimentational and orogenic data from the Pacific region. Even the late Mesozoic operations upon the New Zealand alpine fault were suspended until the Miocene. If this be truly so, the geopolarity stripes are a casualty and must record a series of magnetic events that are independent of the tectonic behaviour of the Earth.

Authoritative opinions on continental margins throughout the world are brought together in the volume edited by C. L. Drake and C. A. Burk (1974) who remark in their introductory paper that 'The exact position of the edge of the continental crust is extremely difficult to determine'; but who make a broad distinction by acoustic differences between thick granite crust and thin oceanic basalt crust. Rare fortunate margins combine the features of both land and sea environments: the Canary Islands, for instance, have ophiolites and appear to mark the transition clearly from continental ocean crust; Cape Verde Islands have carbonatites. Even so, areas like the Bahamas, the Gulf of Mexico, Sea of Okhotsk, and the Sea of Japan still present difficulty and frequently such areas 'show evidence of massive vertical movements'.

Precise location is sometimes masked, too, by sedimentary accumulations 5000–6000 metres thick, and even in these formations hiatuses are common and large. Borehole DSDP 397, west-northwestward of Cape Bojador, West Africa, for instance, showed an absence of deposit between middle Hauterivian and earliest Miocene although the total thickness of sediment present was near 6000 metres. By contrast, the next borehole, 398, off Portugal, gave 'a record deep penetration of 1740 metres sub-bottom, bottoming in Hauterivian limestone and mudstone', yet had but a single hiatus in an otherwise complete stratigraphic succession from Hauterivian almost to the present time. That hiatus was the worldwide mid-Cretaceous break (Horizon B) which appears generally below the Cenomanian in Laurasia and after that stage in Gondwana. Accumulations of this sort, developed under a relatively stable, incessantly subsiding environment, are ideal for seismic reflection

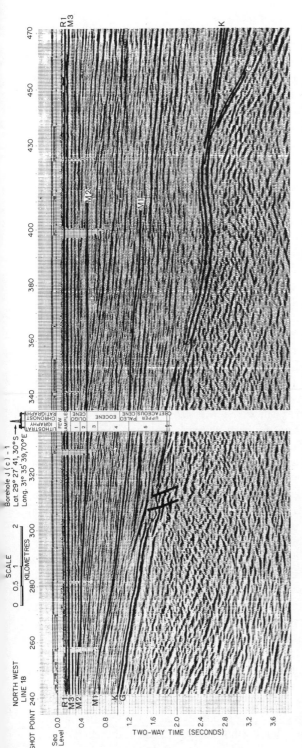

Figure 49 Seismic section and drill log of borehole J(c)-1 24 km offshore from Stanger, Natal, for comparison with the denudational history of the province (Fig. 44) and the geological sections of the coastal plain alongside the lower Umfolozi River (Figs 51-53). The basement of Palaeozoic continental rocks penetrated by the borehole (6, 7) was of Dwyka diamictite and Table Mountain quartzite similar to formations exposed near Stanger in tilted fault-block structure of the Natal Monocline.

Above this basement, and beneath upper Cretaceous marine sediments, lie unconformities corresponding to the Gondwana and Kretacic planations with pronounced seaward dip GG and KK. With successively lessening dip come three early Cenozoic unconformities M1, M2, and M3 representing phases in the development of the Moorland planation which at Uloa, as in the borehole, is the unconformity beneath the lower Miocene (upper Oligocene in the sample log) shelly Pecten Bed. Du Toit and Leith (1974) correctly attributed the slight difference in age in what was evidently a single formation to transgression of the sea over the continental margin. The denudation supplying waste for the Pecten Bed was the Rolling landscape cycle (R1), and its corresponding break in the sequence of sediments lies between the Miocene and Pliocene beds at Uloa (Fig. 53) and elsewhere over the coastal plain.

Intermittent steepening of the older formations relative to the younger in this section indicates periodic tectonic rejuvenations of the Natal Monocline between Jurassic and Recent time, but du Toit and Leith note the absence of compressional folding from the Cretaceo-Cenozoic succession. These authors also noted: 'a prominent set of prograding reflectors and channels are evident over the interval designated middle to upper Eocene.' Horizon A again? (reproduced by permission of SOEKOR) exploration techniques, and occupy sites of maximum vertical subsidence (broad oceanic basins), without underlying subduction structure.

Both subcontinental and suboceanic types of crust meet, of course, beneath the continental margins, and Drake and Kosminskaya (1969) remark upon 'the presence of enormous thicknesses of material in the velocity range 7.6–7.8 km/sec. in the sub-continental type', which might be responsible for changes of elevation at the Earth's surface.

Coastal types

Following Suess, coasts were classified into passive (Atlantic) and active (Pacific) types, but the real difference is that the former are fractured, new coasts produced within Gondwana and Laurasia at the Mesozoic disruption of the supercontinents, whereas the latter were long-established coasts of Gondwana and Laurasia which had advanced into the Pacific by disruptions of the same age.

Monoclinal coasts

Apart from clean fracturing there are: little disturbance of strata, no orogenic mountains, little seismicity, and no marked volcanism along monoclinal coasts. Gravity highs and sedimentary wedges are, on the contrary, common to all passive margins. Epeirogenic tectonics amount to intermittent and repetitive marginal tilting seawards about coastal hingelines that have changed little in position ever since the Mesozoic continental disruptions decided them. Minor block faulting may be accessory, e.g. West Australia, Natal Monocline, and many others.

On the landward side appears a step-like succession of denudational planations, rising inland; and offshore are strata and wedges of sedimentary formations built from the detritus of those denudations and with unconformities corresponding with the stages of planation upon the lands. For the Natal Monocline, seismic studies of the offshore sediments (du Toit and Leith, 1974) (Fig. 49) show that these Cretaceous (and later) sediments are very regularly disposed and that though their original dip towards the basin of the southwest Indian Ocean has intermittently been slightly increased during Cenozoic time with subsidence of that basin, in accord with the fundamental monoclinal tilting of the continental margin, no serious dislocations affect the smooth accumulation of the strata. The few disconformities or unconformities either agree with those in the coastal plain sequence of Cenozoic rocks (q.v.), or are supplementary to the land planations. This latter state probably exists where a later denudation has incorporated or 'swallowed up' the features of a previous intermission, when either the time interval or the altitude difference between them was small (Fig. 49). Both series of events are initiated by the same tectonic episodes. Dating of these events is made by the palaeontology of the marine sequences, and this dating can be transferred directly to the topographic features by means of the unconformities in coastal plains (Figs 51, 52, and 53).

The east and west coasts of Africa, for instance, are passive in this style. They are regarded as wholly within the continental plate. It is they, and especially the Natal Monocline, which furnished the data for table 1. So important is this matter, however, that we cite briefly other examples from other parts of the world. Typically, too, all of these display (in various stages of preservation) a prominent coastal plain of Pliocene age, backed by rising country in which two planation surfaces are evident (a) a rolling Miocene landscape which is incised into (b) an earlier, smooth moorland which was planed over the time interval late-Cretaceous to early Miocene. This moorland surface often bears ancient soil profiles consistent with its prolonged duration — laterite, calcrete, bauxite, clay with flints, and so forth according to the nature of the bedrock (p. 188).

Very similar denudational features appear in the coastal hinterland of Brazil (King, 1962/7) and there is also a precisely similar depositional sequence in the littoral. This identity in geomorphology is carried to great lengths. In both regions the coastal monoclines even change direction in a similar manner from southeast-sloping to south-sloping, and an axis of enhanced uplift runs inland from the angle of directional change. In Africa this axis leads from near Cape Recife northwest to the Compassberg, highest peak in the Cape Province; and in Brazil from Cabo Frio to Pico de Bandeira, the highest peak in that country. The coastal monoclines of Brazil are paralleled inland by two rift valleys — the São Francisco rift which parallels the coast north of Cape Frio and the Paraiba graben which is directed east–west.

The southeast corner of Australia is alike in that the coastal monoclines change direction from southeast-sloping to south-sloping about Cape Howe and the rising axis leads from the angle of change straight to Mount Kosciusko — the highest point of Australia.* The planation surfaces also are like those of the other southern continents in appearance as in their respective ages. This I verified in 1956, although F. A. Craft had previously mapped the cyclic landscapes of New South Wales and Victoria.

India presents a similar set of planations and comparable marginal monoclines showing 'great subsidence and large vertical tectonics' (Demangeot).

Gondwana geologists have been accustomed to find identities of stratigraphy within their respective land masses since the time of Suess, and have explained these agreements very satisfactorily by the hypothesis of continental drift (du Toit, 1937). But the facts adduced here refer to times which postdate the break-up of Gondwanaland and continue right up to the present day. They have nothing to do with the Mesozoic break-up and subsequent continental drift; they refer primarily to Cenozoic tectonic movements operating solely in the vertical sense, both up and down; but as was shown in *The Morphology of the Earth* (King, 1962/7), remarkably similar

* If you need further explanation, try bending a sheet of stiff paper in two monoclines along its east and south edges, and observe the behaviour of the northwest corner.

in timing over the face of the Earth and controlled, with little doubt, by conditions prevailing from time to time in the upper mantle (p. 184).

To complete the picture we consider a couple of examples from the northern hemisphere. The Atlantic slope of the United States displays, from Schooley peneplain time (late Cretaceous to Oligocene) on to the present, an intermittently steepened coastal monocline with associated sequences of offshore sediments (e.g. along the Chesapeake) that agree stratigraphically (Johnson, 1931; King, 1962/7) with the general marginal chronology of the lands within the southern hemisphere. Nor is Britain exempt, for as we have seen (Fig. 46) during the Cenozoic the British Isles have canted progressively eastward, towards the North Sea basin. The history of that basin has been written by Kent (1975) who shows that its site is epicontinental in that Palaeozoic structures are continued beneath it from Britain to Scandinavia. Subsidence began during the Permian (Woodland, 1975), and delta sands accumulated during the middle Jurassic. Tectonic displacements were vertical, and even the Central Graben of the North Sea was not related to opening up of a true ocean basin.

The Cenozoic story may be read on land with the early Cenozoic planation of the Moorland surface (cf. Schooley peneplain), the attitude of which was defined by Linton (Figs 46, 47). Although Linton thought this evident planation and tilting was pre-Cenomanian, I later identified it (King, 1962/7) as the Moorland planation which began in Great Britain in the late Cretaceous and culminated before the Miocene.

Of North Sea development Kent noted: 'The North Sea basin thus shows, in a daily increasing amount of detail, a history of subsidence which is closely similar to that of other so-called "inactive" (Atlantic-type) coasts worldwide'. The features of the North Sea basin have developed without continental drift or any supposed sea-floor spreading. They were controlled wholly by differential vertical tectonics, up in one place, down in another, and on one side at least its margins match with those of major ocean basins, and conform to the researches of this author that '*despite the admitted complications of local factors a single, simple global pattern of tectonic episodes with planational intermissions is recognisable for the whole earth from the mid-Mesozoic until the present day*'. (King, 1962/7, 1976)

Between the two realms of land and sea are monoclinal hinge zones where uplift or subsidence alternate, marine transgressions and regressions occur, and the resulting terrestrial or marine records may alternate. No worldwide eustatic changes of sea level are implied: though these may be independent. Always the landward side goes up, the seaward side goes down. Within this hingeline zone, the regimes of land and sea interfinger and provide a detailed record of tectonic and ensuing events. Only within the zone of hingeline wandering are events reversible. Lateral shifting of the hingeline zone from time to time, either (a) seaward or (b) landward (Fig. 50) produces either: (a) an uplifted shoreline with a coastal plain; or (b) a drowned shoreline,

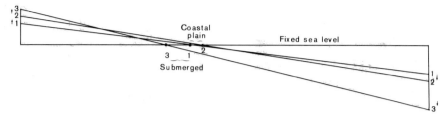

Figure 50 Cross-section of a monoclinal coast, original position 1. The diagram shows that seaward tilting about an axis shifted laterally to position 2 increases the height of the land, deepens the offshore basin and creates a coastal plain from former sea floor. Shifting of the hingeline landward (position 3) increases the height of the land, deepens the offshore basin markedly and causes drowning of the shoreline. The tendency will be for the hingeline, established at the formation of the monocline, to hover about the same position for a long time thereafter. A lateral shift of 20 km is large

respectively. Examples of this type of marginal tectonics are found in all parts of the world.*

The importance of coastal plains

Coastal plains rising to about 180 m at their inner edge are a familiar sight, and the histories that they afford are often similar. Near surface are littoral marine deposits perhaps with Pliocene fossils. These deposits are widely transgressive, often to the inner margin, and they are commonly re-sorted into beach sands and dune sands at stages of the ensuing (Holocene) regression to (and sometimes beyond) the modern shore.

Many examples show the Pliocene deposits to be transgressive on, and unconformable to, earlier Miocene strata; and the bulk of many coastal plain terrains is composed of large wedges of Cretaceous (often late Cretaceous) strata, mostly sandstones and mudstones. A typical example is the coastal plain of Moçambique and Zululand between the Lebombo Mountains and the sea, and continued as a narrow continental shelf beneath the Indian Ocean. The main geomorphic evolution of this coastal plain is clearly expressed in geological sections along the lower Umfolozi and other rivers (Fig. 51). The oldland to the west is composed of rocks belonging to the Karroo System (Palaeo-Mesozoic) ending with the Drakensberg plateau lavas of early Jurassic age. These (Karroo-type) rocks are here flexed down at 15° seaward beneath the coastal plain, and they here constituted the shoreline of Africa after the Jurassic break-up of Gondwanaland and early development of the Natal Monocline. They are exposed in a basalt quarry at Mtubatuba (Fig. 52). In the interior (Lesotho) the basalts are transected by the

* W. C. Pitman (1978), arguing from sedimentary data reached a similar conclusion: 'The most important conclusion of this work is that transgressive or regressive events recorded at numerous subsiding margins may not be indicative of eustatic sea-level rise or fall, respectively, but may be caused by changes in the rate of sea-level change. A decrease in the rate of sea-level rise will be regressive, as will be an increase in the rate of sea-level fall.'

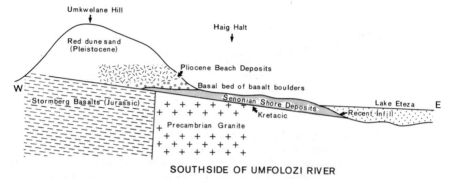

SOUTHSIDE OF UMFOLOZI RIVER

Figure 51 Section of the inner edge of the Zululand coastal plain at Umkwelane Hill on the south side of the Umfolozi River showing relations of the continental basement (Precambrian granite and Stormberg basalts (Jurassic), upper Cretaceous and Pliocene deposits)

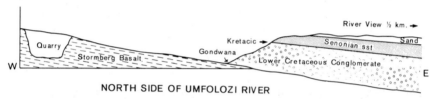

NORTH SIDE OF UMFOLOZI RIVER

Figure 52 Section along the north side of the Umfolozi River at Riverview showing the relationships of the Gondwana and Kretacic planations to the stratigraphy

very smooth (Gondwana) landsurface, and this surface too is apparently flexed down with the basalts and passes beneath lower Cretaceous conglomerate bearing much fossil wood.* At the Moçambique shoreline the basalts are encountered in boreholes at depths of 1800–2500 m below sea level, so the Jurassic monocline is continuous into and beneath the continental shelf. Both the coastlands and the shelf floor also display much Cretaceous block faulting with sliding of the blocks down the monocline towards the sea. This formation of basins is indeed usual upon monoclinal continental shelves in all parts of the world. It often begins during the Permian and is continued intermittently until the mid-Cretaceous. Around East and South Africa (Dingle and Scrutton, 1974) it is succeeded by Aptian, or later, marine transgression (or landward shift of the hingeline of tilt). Kent (1977) also notes that 'the transgression is not eustatic, but has to be tectonic.'

The wood-bearing conglomerates are the earliest deposits to cover the Gondwana landsurface, and farther north in Zululand they are succeeded by marine fossiliferous Aptian, Albian, and Cenomanian strata in succession. These are composed of the detritus shed by the Kretacic planation then developing in the coastal hinterland; as are also 3000 m of continental Cretaceous (Sena Formation) at the inner edge of the Moçambique coastal plain south of the Zambezi where a long trough developed. Strata corresponding

* The actual contact cannot be seen on the Umfolozi River because it lies along a small, lake-filled tributary valley.

207

to the Turonian, and prior to the Campanian, are either absent from, or dubious in Zululand. This hiatus concludes the Kretacic planation which had followed the disruption of central Gondwanaland.

A new regime in the Campanian, involving an uplift of 1000 m or more in the African interior and conversely deep offshore subsidences (monoclinal), brought back the sea over the shore, and widespread upper Cretaceous and Paleocene rocks make huge wedges beneath the coastal plain. Offshore drilling (DSDP 249 and J(c)1) 24 km off the Natal coast (Fig. 49) record both the Turonian interval and the resumption of late Cretaceous sedimentation which in the sea continues even to the Oligocene before this phase of accumulation finished. Farther eastward, however (in the Moçambique basin and over the Moçambique ridge) the Oligocene is generally absent.

This same seismic section (Fig. 49) records also an unconformity in the early Palaeocene, which seemingly has no equivalent planation in the scenery of Natal. If such features formerly existed, they may well have been incorporated in the succeeding Moorland planation.

Over the African hinterland at this same interval of time (late Cretaceous and early Cenozoic) was planed the Moorland cycle, of extreme bevelling and often bearing the well-known laterite duricrust characteristic of that planation.

The long-continued correspondence between terrestrial denudation leading ultimately to mid-Cenozoic planation and duricrusting with the prolonged sedimentation slowing almost to a stop by the mid-Cenozoic, cannot be fortuitous on the scale through which it is developed. The time spans are the same, and the quiescence of the crust is similar for both sets of phenomena. This history, too, is found along the Atlantic seaboard of the United States and southeast Brazil. It is present in Australia. With suitable local digression* it is virtually worldwide in distribution (King, 1962/7).

A fresh tectonic impulse was then necessary to start fresh cycles of landscape development and marginal sedimentation. The impulse came about the early to mid-Miocene and is recorded at Uloa (Fig. 53) on the coastal

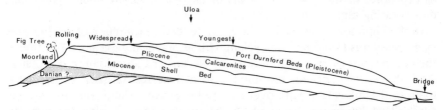

Figure 53 Section north side of Uloa cutting showing four stages of planation (as developed in the nearby landscape of Africa) here represented by unconformities in the stratigraphic sequence of the coastal plain: these are Moorland (early Cenozoic), Rolling (Miocene), Widespread (Pliocene, here the coastal plain), and Youngest (Pleistocene–Recent)

* E.g., farther offshore the new, shelly deposits are often earlier than at the coast, for instance late Oligocene rather than early Miocene. Such transgressional relations are normal.

plain of Zululand 15 km nearer to the sea than Mtubatuba. There, an unconformity with leached Maastrichtian-Danian below is overlain by the pisolitic ironstone base of a Burdigalian *coquina*. The ironstone contained fragments of true laterite, probably derived by erosion of the Moorland duricrust as renewed Miocene denudation got to work developing the Rolling surface in the coastlands. The Miocene shell-bed covered only the eastern one-third of the width of the coastal plain (5 km at the Umfolozi); but though it is only from 3 to 15 metres thick I have followed it along the coastal plain at intervals northwards for 920 km to Maxixe near Inhambane. Over the western two-thirds of the coastal plain denudation occurred and any Miocene sediments are terrestrial, e.g. at Magude.

Far out in the open oceans Miocene sediment of patchy distribution has been recovered from near the crests of certain suboceanic ridges which began to rise about this time, and may be taken to indicate (a) early Miocene tectonic disturbances there, followed by (b) new spreading of waste upon the ocean floors (several phases may be represented). This was interrupted in due course by (c) strong Pliocene cymatogeny which created the modern forms of the ridges.

At the coasts this was marked by renewed seaward tilting and strong transgression so that the Pliocene sea covered the entire width of the coastal plain in Zululand from the sea to the Lebombo oldland at the foot of which maximal marine Pliocene formations are known. At the Umfolozi River for instance, the shoreline advanced from its Miocene position at Uloa for 15 km inland to Umkwelane Hill (Fig. 53 to Fig. 51). The transgressional phase was marked by the laying down of widespread fossiliferous shallow-water calcarenites, much of which was destroyed during the ensuing late Plio-Pleistocene regression when the waste was redistributed in the form of beach and dune deposits marking stages of withdrawal of the sea to and beyond the present coast.

Irregular unconformities lie both above and below these calcarenites and all the Uloa unconformities were recorded photographically by King (1970), as exposures of these weak deposits of fluctuating environments were then deteriorating rapidly under weathering.

In the Umfolozi river section the cross-bedded beach phase of the Pliocene transgression is found on Umkwelane Hill at practically the same position as the beach sands (highly fossiliferous) of the earlier Campanian transgression. And above the Pliocene, red dune sand with terrestrial snail shells makes the top of Umkwelane Hill, over twenty kilometres inland from the present coast. Two kilometres seaward of the present shore, rising through 100 metres of water, the last of the dune lines of the late Pliocene retreat now make an offshore reef. So a late return of the sea (post-Pliocene) is responsible for the drowned river mouths and lagoons which are locked in behind the current line of backshore dunes, some of which reach immediately to 150 metres above the sea.

Following the great monoclinal tiltings of the continental margin towards the sea, the rivers of the mainland have been entrenched by 350–550 metres. These entrenchments are continued (by turbidity current action) across the shelf, and submarine canyons are numerous along the edge of southeast Africa. From some of these (Tugela, Zambezi) large cones of debris have been spread. The latter carries coarse gravel of granitic and metamorphic rocks 800 km from the river mouth to near southern Madagascar.

Seismic reflection exercises conducted under the guidance of Soekor (Fig. 49), show an amazingly regular and little-disturbed sequence of Cretaceous to modern sedimentation offshore of the Zululand coastal plain out to the Agulhas fracture zone which marks the original outline of South Africa.

The geomorphic history of the Zululand–Moçambique coastal plain, embracing alternate phases of terrestrial and marine activity through Cretaceous and later time, has been stated at length because it is typical of many continental margins. As has been remarked above (Fig. 50) such coastal plains are hinge zones (Fig. 50) between intermittently uplifted continents and intermittent subsidences of the ocean floor surrounding those continents. Events within the marginal hinge zone interfinger from both sides. Nonetheless, the hinge zone itself may remain relatively stable (horizontally) or narrow. That this is so in Natal is shown by the strong coincidence between the late Cretaceous and modern shorelines of the Province.

Many observers have explained the marginal transgressions and regressions of the sea eustatically. This explanation is simple, and it may be satisfactory of some instances; but it is necessarily of worldwide application and fails to explain nuances of contrast observed from place to place that are a natural result of slight changes of direction between shoreline and hingeline. Nor does eustasy explain the repeated, long-continuing monoclinal uplifts of the land combined with subsidence of the ocean floor (both of which contrasted movements may be of thousands of metres), nor of the contemporaneous relations between them.

I have naturally chosen to deal with a coastline familiar to me and that I have studied for many years; but the summation of the East African coast by Kent (1974) is not dissimilar: This coast came into existence during the mid-Jurassic, and 'the continental edge (not the shore) has been progressively built out during the deposition of some tens of thousands of feet of sediments.' 'The movements documented by geological and geophysical evidence were essentially vertical, and the area is characterized by vertical faults. There is no evidence of low-angled faults or other features which would be associated with plastic stretching of the crust. Lastly, there was a major regression in the Oligocene, locally associated with faulting, and lower Miocene is widely transgressive. Thick (10,000–12,000 m) later Miocene and Pliocene are known from seismic evidence in synclinal belts.' The general resemblance with southeast Africa is clear; and much of the late Cenozoic heavy sedimentation can be attributed to the well-known, more active tectonism of East Africa at that time.

The coasts of Angola, West Africa, Brazil, India are all similar, and all can be used to link tectonically and in time the features of the lands and the ocean basins. The Brazilian coastline, for instance, is monoclinal and on the lands the same set of planations have been found in its hinterland as are present in Natal. I mapped them myself in 1954 (King, 1962/7). The sedimentation there also matches that about Africa in its main features. Kowsmann *et al.* (1977) record prograded wedges of Cretaceous formations (with similar unconformities and a transgressive lower Miocene as the most defined features). There are upper Miocene shales. Coarse Pliocene and Pleistocene deposits testify to strong uplift of the nearby land. Extensive Holocene sands, silts, and clays show regression of the shoreline.

Further evolution of monoclinal coasts

Certain monoclinal coasts seem to be inherently unstable and early in their evolution develop basins along the continental shelf. These may be sited in accordance with pre-existing structures, e.g. around the southern end of Africa where the basins and ridges conform with the major structures of the Cape Fold Mountain Belt; but usually they appear to be random, as do three Cretaceous basins on the western aspect of Africa south of the Walvis Ridge.

On the opposite side of the South Atlantic fresh basins were formed where four pre-Cretaceous basins, consonant with the Gondwana geology of the Argentine, extend into the broad continental shelf. These are Cretaceous and younger, and have different geographical distribution like the shelf basins of southwest Africa. The earliest 'Atlantic' sediments on both coasts (Brazil and Angola) are lagoonal Aptian and marine Albian, following upon earliest Cretaceous basalt lavas (exceeding 1500 metres in Paraná) and in the Kaokoveld.

But the Atlantic coast of North America is richest of all in these coastal basins, nine of them strung out along the offshore zone from the Labrador shelf to the Gulf of Mexico. Most are grabens in the basement, and the thickness of Mesozoic and Cenozoic sediments within them may measure between 7 km and 14 km.

Sometimes the mass and depth of sediments does not account isostatically for the amount of basin-floor sinking, and independent crustal floor sinking needs to be postulated. Great thicknesses of shallow-water type sediments are characteristic off the continental shelf edge. The bulk of the eugeosynclinal sequence is turbidites trapped in a trench rather than pelagic deposits of the open ocean.

Off the eastern United States, basin formation upon the shelf is sometimes aided by *rise* of the shelf edge between the mio- and eugeosynclines. Rise of the shelf edge may be accompanied by yet deeper sinking of the eugeosyncline (see perhaps on Fig. 48). And the structure tends more and more towards the features of subductive coasts. Thus beneath the Labrador shelf are several faults stepping down seawards, and beneath the slope a great fault dropped the sea floor. The seaward section shows the mantle here *plunging down* against the fault; thickened anomalous mantle above it also plunges. Even the oceanic basalt layer dips towards the fault and a deep basin filled with sediment (2.5–2.6 km/s) represents the usual position of the continental rise. There is no topographic rise here, but all the elements for a geosyncline are present and subduction appears to be developing.

Towards coasts of Pacific type

Other sections give similar elements and distributions, notably the Grand Banks and the main coast from Nova Scotia southward to Cape Hatteras. Throughout, the most striking feature is the downward plunge of the mantle zone and oceanic crust against the great fault which outlines the continent. This drastic subsidence has developed since the Mesozoic fracturing of the continental edge. So here a typical 'passive' type of coast has developed a similar type of subductive structure as the 'active' circum-Pacific type, except that the final, vertical uplift has not yet taken place to raise new mountain ranges upon the continental margin.

At the Gulf of Mexico 'vertical movements have predominated since the Palaeozoic and large thicknesses of sediment (15 km) have accumulated, depressing the floor still further and engendering salt diapiric structures.' In the Caribbean, too, though its structures (within island arcs) are more Pacific in type, data indicate a Cenozoic and late Mesozoic history of predominantly vertical and strike-slip movements as being responsible for the marginal deformation. Monoclinally steepening seaward, and with wedges of sediment offshore, the coasts have been rejuvenated several times in the Cenozoic, and in their most extreme evolution they develop miogeosynclines upon the shelf area and eugeosynclines at the foot of the continental slope, but modern instances involving orogenesis and the elevation of marginal geosynclines into new lands are few. New Zealand's Cretaceous Rangitata Orogeny is perhaps an example, reactivated in the later Kaikoura Orogeny (Plio-Pleistocene).

Evolution of ancient complex coasts (Pacific type), or: What happened to the former coastlines of Gondwana and Laurasia?

The Pacific-type coasts are immensely older than the simple monoclinal coasts, and formed the evolving margins of Gondwana and Laurasia during (possibly) Proterozoic, Palaeozoic, and early Mesozoic time. Those parts which still survive today face the Pacific Ocean, where they have lived out the Cenozoic chapter of their history — nearly all of them in zones of nearby subduction. This gives reason to believe that subduction is a more advanced mode of marginal evolution than is found upon monoclinal coasts. Indeed, we have seen that off the northeastern coasts of North America geosynclinal sedimentation and subduction have recently begun, and conceivably someday will lead to orogenesis.

The eastern side of Australia was formerly one land with New Zealand. The Tasman Sea did not open until 60–80 Ma ago. Prior to that the Ordovician geology of Preservation Inlet corresponds with the graptolite zones of Bendigo, although those latter are five times as thick. And the marine Triassic of New Zealand becomes non-marine and resembles that of eastern Australia.

The subduction zone from Tonga passes through New Zealand to the Solander Trough and Australia is exempt from the subduction phenomenon.

But the complex Palaeozoic orogenic structures of eastern Australia are very different from the flat-lying Karroo beds of Natal, so the response of the former to Cenozoic monoclinal tilting (Ollier, 1978) is somewhat more complex than appears in Natal (King, 1972).

Uniformity does not exist between various parts of the circum-Pacific orogenic structures; but most are mountainous in the hinterland, and the mountains are orogenic and built of the contents of Cenozoic geosynclines. Modern island arcs and deep trenches are abundant, and eastern Asia is garlanded by youthfully enlarged marginal seas between the island arcs and the mainland. This type of geography has existed many times during the late Cretaceous and Cenozoic, but again and again during that timespan orogenesis has supervened and raised the several zones of structure up into mountain ranges complete with volcanic piles. And each time the mountains have been accreted (perhaps with continental drift) to the nearby continent.

The major episodes of orogenesis are usually synchronous with the episodes of elevation on the monoclinal coasts: late Cretaceous, early Miocene, early Pliocene, and Pleistocene. This argues strongly for global control of tectonic phenomena, perhaps through mantle activity. Progressively older trench slopes first subside, and then are uplifted while being intruded by granitic plutons and incorporated into the continent.

A question which arises here is: how are the mountains so hugely uplifted? The altiplano of the Andes, for instance, was planed at relatively low levels during the Pliocene and was carried up to 4000 metres with volcanic cones which surmount it to 7000 metres at the same time that the offshore Chile trench subsided to minus 8000 metres. The link between them is clearly the Benioff Zone descending through hundreds of kilometres, deep in the mantle zone. That the mechanism of uplift is by differential vertical laminar tectonics is reasonable; but what activates the differential vertical laminar tectonics? On seismic data Molnar and Oliver (1969) consider that hot mantle may be present beneath the altiplano of northern Chile and southeastern Peru where the broad uplifted region is inland of the zone of active volcanoes. If that mantle were not only hot, but charged abundantly with volatiles, we arrive at the familar pattern of crustal orogenic evolution by vertical tectonics — powered by the de-gassing of the mantle.

But maps of the Pacific emphasize the contrast between the marginal features of eastern Asia and the straight steepness of the Americas' front to the Pacific with only a narrow continental shelf and no marginal seas. Any former island arcs are there accreted to the mainlands. These fundamental differences have already been ascribed to the pre-Cretaceous westerly drift of both the parent supercontinents producing a 'trailing edge' upon eastern Asia, while both North and South America swept up any such arcuate features and incorporated their structures into its 'advancing edges', where they may now be inspected as orogenic structures. Cenozoic structures formed by vertical tectonics still follow the inherited patterns.

Yet the western margins of North and South America differ profoundly for the pre-Cretaceous palaeogeography of Laurasia and of Gondwana differed from each other. In the north, Alaska and the Aleutian structures belong more to Asia than they do to North America, e.g. geophysical evidence, indeed, continues the Brooks Range structure across the Chukchi shelf to the Chukchi-Anadyr fold belt of far east Siberia, and the Cenozoic folding across the shelf produced a dry-land connection between the continents until about a million years ago when the low mountains were eroded away. The northerly extent of the American continental shelf (which includes the Queen Charlotte Basin north of Vancouver Island) is of Eocene volcanics forming part of an early Cenozoic island arc overlain by 2–3 km of Miocene and Pliocene fill.

Southward there are *no* offshore trenches along the whole Pacific coast of the United States. These features reappear only off Central America whence they continue the length of South America to latitude 35°S. Seismic studies agree. As was pointed out in 1964 by Girdler, there are profound seismic differences between the western margins of the Americas. Whereas South America was one of the first places from which deep-focus earthquake shocks (between 100 and 700 km depth) were recorded, none from even 100 km are known from western North America. The difference of deep structure between them would appear to be that western North America is underlain by an arrested continuation of the East Pacific Ridge, which has a low-density mass of anomalous mantle at shallow depth as is usual with submarine ridges, and which is typified by mid- and late Cenozoic uplifts amid the western cordillera. Central America and Andean South America, on the contrary, appear from the seismic and gravimetric data to be located over a subductive slab-like structure dipping eastward beneath the continent.

The vertical movements are modern and their magnitude large. From the Altiplano of the Andes to the floor of the Peru–Chile trench (both of them probably of Pliocene date) tectonic rejuvenations measure a difference of 9900 m vertically. This measure is accomplished in a zonal width of 120 km, by monoclinal tilting. Lozenge-shaped relics of the Pliocene planation upon the west Andean spurs south of Santiago were inspected from the air and found to be noticeably tilted.

Where Gondwana and Laurasia collided (p. 6) along the line from

Gibraltar to Burma, those lengths of opposed marginal structures are now to be seen in the back-to-back structured belt of magnificent mountains that span a zone 12,000 km long from west to east. In them too the relicts of Palaeozoic geosynclines may be discerned.

Epilogue

Looking back over the writing of this book, I am struck by the differences that have emerged between the *Geophysicist's Credo* with which it began, and what I would be prepared to believe now.

During a long career in geomorphology and engineering geology, I have sometimes encountered problems upon which there was a woeful lack of data but where a responsible decision had to be made, perhaps rapidly because of emergency. Such instances arise where landslides are incipient, earthquakes or volcanic eruptions occur, or even by stress of abnormal weather. At Westville where I write, on 30 December 1977 between midnight and 2.30 am, 250 mm of rain fell (in Durban, 10 km away, there was no rain). Then between 9 pm and midnight of the same day a further 260 mm of rain deluged. (Over 20 inches of rain, in two spells between midnight and midnight!) This is how geology works, in short violent spasms followed by prolonged quiet intermissions.

One needs to be chary of theories built upon assumptions that geological processes act slowly and accumulatively to give predictable results that may even be computerized by programming. They do nothing of the sort; stresses may build up in rock masses for centuries, but the resulting rock burst in a deep mine takes place in a fraction of a second. Other Earth processes may be equally sporadic and violent.

Dr Henry Olivier, the brilliant Gondwana engineer who designed or built several of the world's largest dams (and to whom a foundation failure was no more tolerable than a structural one) was familiar with this kind of situation: 'One knows exactly the properties of copper and steel. How can anyone ever know the properties of rock, which is not a homogeneous material, from one metre to another? . . . Technical know-how has to be reinforced by cumulative experience and precedents.'

In the final chapter of his book *Damit*, Olivier sums up this pioneering forefront as 'judgement engineering' and 'judgement geology'. Judgement geology may be defined as the stating of geological truth beyond the possibility of direct observation or by derivation from already established data. The judgement may, of course, be verified (or negated) by information

215

subsequently acquired, but in the beginning it is largely *intuitive* — the subconscious result of much varied experience.

Geology is at present in danger. Danger of too-great emphasis upon physical instruments, with reduced reliance upon personal experience of Earth phenomena. Computers and other gadgets are marvellous; but they give, by mathematics, a 'majority opinion', which cannot match the originality of a single gifted brain. A true example: computer EVM 'M-20' had been programmed to play chess. This it did logically and well. No ordinary player could take liberties with it and survive. But logicality was its limitation, and one day it played Grandmaster David Bronstein. After black's 13th move the master declared a mate in 10! He had taken the measure of the computer, it would not understand the sacrifice of pieces for time or for position. For the lesson which this game affords the score is given here. Please enjoy it as I did — by courtesy of Grandmaster Bronstein.

White: D. Bronstein   Black: EVM 'M-20'

1 P–K4	P–K4	14 N×NP+	
2 P–KB4	P×P	14	Q×N
3 N–KB3	N–KB3	15 N×QBP++	
4 P–K5	N–N5	15	K–K2
5 P–Q4	P–KN4	16 N–Q5+	K–K3
6 N–QB3	N–K6	17 N×P++	
7 Q–K2	N×B	17	K–K2
8 N–K4	N–K6	18 N–Q5+	K–K1
9 N–B6+	K–K2	19 Q×QB+	
10 B–Q2	N× BP+	19	Q–Q1
11 K–B2	N×R	20 N–B7+	K–K2
12 N–Q5+	K–K3	21 B–N4+	P–Q3
13 Q–B4	P–N4	22 B×P+	Q×B
Here Bronstein declared		23 Q–K8 mate	
'Mate in ten'			

Nowadays, since World War II, tremendous advances in the gathering of factual data on the nature of the Earth's crust and mantle, the bathymetry and physical properties of the ocean floor, and global tectonics have presented students of the Earth with a bewildering mass of information requiring systematization and synthesis. The time is ripe for bold judgements — by minds well steeped in the information and seeking coordination of data from place to place and discipline to discipline. Most of the 'models' that have been put forward were founded on researches in the North Atlantic: with respect it is submitted that a change of venue is now desirable. The five widely-dispersed Gondwana continents and the several austral oceans afford a richer, wider field for new comparative studies than the North Atlantic and its margins can ever do.

And if a further reason be required, it would surely be that given by G. H. Hardy in *A Mathematician's Apology*: 'It is never worth a first class man's time to express a majority opinion. By definition there are plenty of others to do that'.

References

Adams, C. J., 1980, *Correlations of Precambrian and Palaeozoic orogens in New Zealand, Marie Byrd Land (West Antarctica), Northern Victoria Land (East Antarctica) and Tasmania Gondwana*, 5, 191–7.

Adkin, G. L., 1949, The Tararua Range as a unit of the geological structure of New Zealand. *R. Soc. NZ, Rep. 6th Sci. Congr.*, pp. 260–72.

Alvarez, W., 1973, The application of plate tectonics to the Mediterranean region, in *Implications of Continental Drift to the Earth Sciences*, Academic Press, London, pp. 901–8.

Alvarez, W., Cocozza, T., and Wezel, F. C., 1974, Fragmentation of the Alpine orogenic belt by microplate dispersal, *Nature*, **248**, 309–14.

van Andel, T. H., 1968, The structure and development of rifted mid-ocean rises, *J. Mar. Res.*, **26**, No. 2, 144–61.

van Andel, T. H. and Bukry, D., 1973, Basement ages and basement depths in the eastern equatorial Pacific from Deep Sea Drilling Project Legs 5, 8, 9 and 16, *Geol. Soc. Am. Bull.*, **84**, 2461–9.

Athavale, R. N., 1973, The Indian Plate and the Himalayas, in *Implications of Continental Drift to the Earth Sciences*, Academic Press, London, pp. 177–30.

Atwater, T., 1970, Implication of plate tectonics for the Cenozoic tectonic evolution of western North America, *Geol. Soc. Am. Bull.*, **81**, 3513–86.

Baker, B. H., 1965, An outline of the geology of the Kenya rift valley, in *Rep. Geol. Geophys. of the East African Rift Valley System*.

Baker, B. H., 1971, Explanatory note on the structure of the southern part of the African rift system, in *Tectonics of Africa*, UNESCO, Paris, pp. 543–8.

Bardet, M. C., 1973, Metakimberlites, *Int. Conf. of Kimberlites, University of Cape Town*, Extended Abstracts, pp. 15–17.

Belderson, R. H., Kenyon, A. H., Stride, A. H., and Stubbs, A. R., 1972, *Sonographs of the Sea Floor*, Elsevier, Amsterdam, 185 pp.

Beloussov, V. V., 1970, Against the hypothesis of ocean-floor spreading, *Tectonophys.*, **9**, 489–511.

Bergh, H. W., and Norton, I. O., 1976, Prince Edward fracture zone, *J. Geophys. Res.*, **81**, 5221–39.

Bercheimer, H., 1969, Direct evidence for the composition of the lower crust and the Moho, *Tectonophys.*, **8**, 97–105.

Berner, H., Ramberg, H., and Stephansson, O., 1972, Diapirism in theory and experiment, *Tectonophys.*, **15**, 197–218.

Bernoulli, D., and Jenkyns, H. C., 1974, Alpine, Mediterranean and central Atlantic Mesozoic facies in relation to the early evolution of the Tethys, in *Modern and*

219

Ancient Geosynclinal Sedimentation: Soc. Econ. Palae. & Min. Spec. Publ. No. 19.

Birot, F., and Dresch, J., 1974, Les Surfaces d'applanissement dans les Regions Mediterranean, *UNESCO Symp. on Planation Surfaces, Leningrad, May*, 1974.

Björnsson, S. (ed.), 1967, Iceland and mid ocean ridges: report of a symposium, *Visindafelag Islendinga*, **38.**

Bodvarsson, G., and Walker, G. P. L., 1964, Crustal drift in Iceland, *Geophys. J.*, **8,** No. 3, 285–300.

Bond, G., and Bromley, K., 1970, Sediments with the remains of Dinosaurs near Gokwe, Rhodesia, *Palaeogeogr. Palaeoclimat. Palaeoecol.*, **8,** 313–27.

Bott, M. H. R., 1972, The evolution of the Atlantic north of the Faeroe Islands, *Implications of Continental Drift to the Earth Sciences*, Academic Press, London, pp. 175–87.

Brookfield, M. E., 1977, The emplacement of giant ophiolite nappes, I. Mesozoic–Cenozoic examples, *Tectonophys.*, **37,** 247–303.

Bullard, E. C., 1968, Reversals of the earth's magnetic field: The Bakerian Lecture 1967, *Phil. Trans. R. Soc. Lond.*, A, **263,** No. 1143, 481–524.

Bullard, E. C., 1975, Overview of plate tectonics, in *Petroleum and Global Tectonics Princeton*, pp. 5–52.

Burke, K., and Whiteman, A. S., 1972, Uplift, rifting and the break-up of Africa, in *Implications of Continental Drift to the Earth Sciences*, Academic Press, London, pp. 735–55.

Butler, B., 1959, Periodic phenomena in landscapes as a basis for soil studies, *CSIRO Austr. Soil. Publ.*, No. 14.

Cairé, A., 1971, Chaines Alpines de la Mediterranée Central, in *Tectonics of Africa*, UNESCO, Paris.

Carey, S. W., 1976, *The Expanding Earth*, Elsevier, Amsterdam.

Carey, S. W., 1978, A Philosophy of the Earth and Universe, Johnson Lecture.

Carr, J. B., 1968, When did mantle convection cease? *Tectonophys.*, **6,** No. 5, 413–24.

Carter, D. J., Audley-Charles, M. G., and Barber, A. J., 1976, Stratigraphical analysis of island-arc–continental margin collision in eastern Indonesia, *J. Geol. Soc. Lond.*, **132,** 179–98.

Cloos, H., 1955, Geologische Strukturkarte der Mittelgebirge, *Geol. Rund.*, **44,** 480.

Cochran, J. R., 1973, Gravity and Magnetic Investigations in the Guiana Basin, Western Equatorial Atlantic, *Geol. Soc. Am. Bull.*, **84,** pp. 3249–68.

Cohen, C. R., Schamel, S., and Boyd-Kaygi, 1980, Neogene deformation in northern Tunisia: origin of the eastern Atlas by microplate – continental margin collision, *Geol. Soc. Am.*, **91.**

Coleman, R. C., 1974, Geologic background of the Red Sea, in *The Geology of Continental Margins*, pp. 743–51.

Cook. K. L., 1969, Active rift system in the Basin and Range Province, *Tectonophys.*, **8,** 469–513.

Cox, A., 1973, *Plate Tectonics and Geology*, W. H. Freeman, Reading, pp. 529.

Craddock, C., and Hollister, C. D., 1976, Geological Evolution of the Southeast Pacific Basin, *DSDP leg 35*, **35,** Washington.

Craft, F. A., (193), The physiography of the Shoalhaven River Valley, Parts I–VI, *Proc. Linn. Soc. NSW*, **56** and **57.**

Craft, F. A., 1933a, The surface history of Monaro, New South Wales, *Proc. Linn. Soc. NSW*, **58,** 229–44.

Craft, F. A., 1933b, The coastal tablelands and streams of New South Wales, *Proc. Linn. Soc. NSW*, **58,** 437–60.

Curray, J. R., 1965, Structure of the continental margin off central California, *NY Acad. Sci. Trans.*, Ser. II, **27,** 794–801. 1970–72, *R. Soc. Tasmania*, **112,** pp. 5–19.

DSDP, Reports on Legs 13 (Mediterranean) 15, (Caribbean) 16, 17, *Geotimes*, **15, 16, 17**.

Dalziel, I. W. D., 1970, Large scale folding in the Scotia Arc, in *Antarctic Geology and Geophysics*, Oslo, pp. 47–55.

Dalziel, I. W. D., Dott, R. H., Winn, R. D., and Bruhn, R. L., 1975, Tectonic relations of South Georgia Island to the southernmost Andes, *Geol. Soc. Am. Bull.*, **86**, 1034–40.

Dalziel, I. W. D., and Elliot, 1971, Evolution of the Scotia Arc, *Nature*, **233**, 246–52.

Davies, H. L., and Smith, I. E., 1971, Geology of eastern Papua, *Geol. Soc. Am. Bull.*, **82**, 3299–312.

Demangeot, J., 1975, Recherches geomorphologiques en Inde du Sud, *Z. Geomorph.*, **19**, 229–72.

Demenitzkaya, R. M., and Karasik, A. M., 1969, The active rift system of the Arctic Ocean, *Tectonophys.*, **8**, 345–51.

Deuser, W. G., 1970, Hypothesis of the formation of the Scotia and Caribbean Seas, *Tectonophys.*, **10**, 391–402.

Dewey, J. F., and Bird, B. J. M., 1970, Mountains and belts and the new global tectonics, *J. Geol. Soc. Lond.*, **130**, 183–204.

Dewey, J. F., and Bird, B. J. M., 1971, Origin and emplacement of the ophiolite, *J. Geophys. Res.*, **76**, 3179–206.

Dewey, J. F., and Kidd, W. S. F., 1977, Geometry of plate accretion, *Geol. Soc. Am. Bull.*, **88**, 960–8.

Dewey, J. F., Pitman, W. C., Ryan, W. B., and Bonnin, J., 1973, Plate tectonics of the Alpine system, *Geol. Soc. Am. Bull.*, **84**, 3137–80.

Dingle, R. V., and Scrutton, R. A., 1974, Continental break-up and the development of Post-Palaeozoic sedimentary basins around southern Africa, *Geol. Soc. Am. Bull.*, **85**, 1467–79.

Drake, C. L., and Burk, C. A. (Eds), 1974, *The Geology of Continental Margins*, Springer Verlag.

Drake, C. L., and Girdler, R. W., 1964, A geophysical study of the Red Sea, *Geophys. J. R. Astron. Soc.*, **8**, 473–95.

Drake, C. L., and Kosminskaya, 1969, The transition from continental to oceanic crust, *Tectonophys.*, **7**, Nos 5–6, 363–84.

Eardley, A. J., 1962, *Structural Geology of North America*, 2nd edn, New York, 624 pp.

Egyed, L., 1963, The expanding earth? *Nature*, **197**, 1059–60.

Einarsson, T., 1949, *The Eruption of Hekla 1947–1948*, Parts IV, 2.3, Icelandic Scientific Society, Reykjavik.

Einarsson, T., 1951, *The Eruption of Hekla 1947–1948*, Part V, 2, Icelandic Scientific Society, Reykjavik.

Einarsson, T., 1967, The Icelandic fracture system and the inferred causal stress field, *Visindafelag Islendinga*, **38**, 128–41.

Einarsson, T., 1968, Submarine ridges as an effect of stress fields, *J. Geophys. Res.*, **73**, 7561–76.

Einarsson, T., 1975, Several Problems in Radiometric Dating, *Jökull*, **25**, 15–33.

Einarsson, T., 1976, Upper Pleistocene volcanism and tectonism in the southern part of the median active zone of Iceland, *Problems in Geology and Geophysics*, Part I, Soc. Sci. Iceland, *Greinar*, V, 119–59.

Einarsson, T., 1977a, The Upper Tertiary beginning of zonal volcanism and tectonism in Iceland, *Greinar*, VI, 24–49.

Einarsson, T., 1977b, Palaeomagnetism and the possible effect of depth of burial and non-hydrostatic stresses, *Greinar*, VI, 66–74.

Ewing, J. I., and Ewing, M. 1959, Seismic refraction measurements in the Atlantic

Ocean basins, in the Mediterranean Sea, on the mid-Atlantic ridge and in the Norwegian Sea, *Geol. Soc. Am. Bull.*, **70**, 291–318.

Ewing, J. I., and Ewing, M., 1967, Sediment distribution on the mid ocean ridges with respect to spreading of the sea floor, *Science*, **156**, No. 3782, 1590–2.

Ewing, J. I., Talwani, M., Ewing, M., and Edgar, T., 1967, Sediments of the Caribbean, *Studies in Tropical Oceanography*, Vol. 5, Miami University, pp. 88–102.

Ewing, J. I., Ewing, M., Aitken, T., and Ludwig, W. J., 1968, *North Pacific Sediment Layers Measured by Seismic Profiling*: Am. Geophys. Un. Monogr. No. 12, pp. 147–73.

Ewing, M., 1965, The sediments of the Argentine Basin, *Q. J. R. Astron. Soc.*, **6**, 20–7.

Ewing, M., Eittreim, S., Truchan, M., and Ewing, J. I., 1969, Sediment distribution in the Indian Ocean, *Deep Sea Res.*, **16**, 231–48.

Ewing, M., Ewing, J., and Talwani, M., 1964, Sediment Distribution in the Oceans, *Geol. Soc. Am. Bull.*, **56**, 17–36.

Fairhead, J. D., and Girdler, R. W., 1971, The seismicity of Africa, *Geophys. J. R. Astron. Soc.*, **24**, 271–301.

Falvey, D. A., 1974, The development of continental margins in plate tectonic theory, *Aust. Pet. Explor. Ass. J.*, 95–106.

Ferm, J. C., 1974, Carboniferous palaeogeography and continental drift, *C.R. 7th Int. Congr. on Stratigraphy and Geology of the Carboniferous Krefeld*, Vol 3, pp. 9–25.

Fischer, A. G., 1975, Origin and growth of basins, in *Petroleum and Global Tectonics*, Princeton University Press.

Fischer, R. L., and Hess, H. H., 1963, Trenches, in *The Sea*, Vol. 3, Wiley, New York, pp. 411–36.

Fitch, T. J., and Scholz, C. H., 1971, Mechanism of underthrusting in southwest Japan: a model of convergent plate interactions, *J. Geophys. Res.*, **76**, 7260–92.

Florensov, N. A., 1969, Rifts of the Baikal Mountain Region, *Tectonophys.*, **8**, 443–56.

Flores, G., 1970, Suggested origin of the Moçambique Channel, *Geol. Soc. S. Africa Trans.*, **73**, 1–16.

Fox, P. J., Schreiber, E., and Petersen, J. J., 1973, The geology of the oceanic crust: compressional wave velocities of oceanic rocks, *J. Geophys. Res.*, **78**, 5155–72.

Gansser, A., 1973, Facts and Theories on the Andes, *J. Geol. Soc. Lond.*, **129**, 93–131.

Gansser, A., 1974, Himalaya, in A. M. Spencer (Ed.), *Mesozoic–Cenozoic Orogenic Belts*, Geol. Soc. Lond., Spec. Publ. No. 4, pp.

Gardner, J. V., 1970, Submarine Geology of the Western Coral Sea, *Geol. Soc. Am. Bull.*, **81**, 2599–614.

Gass, I., and Gibson, I. L., 1969, Structural evolution of the rift zones in the Middle East, *Nature*, **221**, 926–30.

George, T. N., 1966, Geomorphic evolution in Hebridean Scotland, *Scot. J. Geol.*, **2**, 1–33.

Gerth, H., 1955, *Der geologische Bau der sudamerikanischen Kordillere*, Borntraeger, Berlin.

Green, A. G., 1972, Seafloor spreading in the Maçambique Channel, *Nature, Phys. Sci.*, **236**, 19–21.

Green, H. W., 1974, A CO_2 charged asthenosphere, *Nature*, **238**, No. 79, 2–5.

Grindley, G. W., 1974, New Zealand, in A. M. Spencer (Ed.), *Mesozoic–Cainozoic Orogenic Belts*, Geol. Soc. Lond., Spec. Publ. No. 4, pp. 387–416.

Gunn, R., 1949, Isostasy — Extended, *J. Geol.*, **57**, 263–79.

Hailwood, E. A., 1977, Configuration of the magnetic field in early Tertiary times, *J. Geol. Soc. Lond.*, **133**, 23–36.

Hamilton, W., 1973, Tectonics of the Indonesian Region, *Implications of Continental Drift*, **2**, 879–84.

Hamlyn, P. R., and Bonatti, E., 1980, Petrology of Mantle-derived Ultrabasics from the Owen Fracture Zone, *Earth Planet. Sci. Lett.*, **48**,

Harrington, H. J., Wood, B. L., McKellar, I. C., and Lessen, G. J., 1965, The geology of Cape Hallet-Tucker Glacier District, in *Antarctic Geology*, Amsterdam, pp. 220–8.

Hast, N., 1969, The state of stress in the upper part of the Earth's crust, *Tectonophys.*, **8**, 169–210.

Hayes, D. E., and Pitman, W. C., 1972, Review of Marine Geophysical Observations in the Southern Ocean, in *Antarctic Geology & Geophysics, IUGS*, Sec. B, No. 1, 725–32.

Heezen, B. C., 1969, The World Rift System, *Tectonophys.*, **8**, 269–79.

Heezen, B. C., and Bunce, E. T., 1964, Chain and Romanche fracture zones, *Deep Sea Res.*, **11**, 11–33.

Heezen, B., and Fornari, D. J., 1976, *Initial Reports of the Deep Sea Drilling Project*, Vol. 30, pp.

Heezen, B., and MacGregor, I. D., 1973, The evolution of the Pacific, *Sci. Am.*, November, pp. 102–12.

Heirtzler, J. R., 1973, The evolution of the North Atlantic Ocean, *Implications of Continental Drift to the Earth Sciences*, Academic Press, New York, pp. 191–6.

Heirtzler, J. R., le Pichon, X., and Baron, J. G., 1966, Magnetic anomalies over Reykjanes Ridge, *Deep Sea Res.*, **13**, 427–43.

Heirtzler, J. R., *et al.*, 1968, Marine magnetic anomalies, geomagnetic field reversals, and motions of the ocean floor and continents, *J. Geophys. Res.*, **73**, 2119–36.

Hermes, J. J., 1968, The Papuan geosyncline and the concept of geosynclines, *Geol. en Mijnb.*, **47**, 81–97.

Herron, E. M., 1972, Sea-floor spreading and the Cainozoic history of the east-central Pacific, *Geol. Soc. Am. Bull.*, **83**, 1671–91.

Herron, E. M., Dewey, J. F., and Pitman, W. C., 1974, Plate tectonics model for the evolution of the Arctic, *Geology*, **2**, 377–80.

Hess, H. H., 1966, The evolution of ocean basins, in M. N. Hill (Ed.), *The Sea, Ideas and Observations*, Interscience, New York.

Hey, R., Johnson, G. L., and Lowrie, A., 1977, Recent plate motions in the Galapagos area, *Geol. Soc. Am. Bull*, **88**, 1385–403.

Hilde, T. W. C., Uyeda, S., and Kroenke, D., 1977, Evolution of the western Pacific and its margin, *Tectonophys.*, **38**, 145–6.

Houtz, R., Ewing, J., and Embley, R., 1972, Profiler data from the Macquarie Ridge Area, *Am. Geophys. Un., Antarctic Res. Ser.*, **15**, pp. 239–47.

Hsu, K. J., 1972, When the Mediterranean dried up, *Sci.Am.*, December, 27–36.

Illies, J. H., 1969, An intercontinental belt of the world rift system, *Tectonophys.*, **8**, 5–29.

Irving, E., and Couillard, R. W., 1973, Cretaceous normal polarity interval, *Nature, Phys. Sci.*, **244**, No. 131, 10–11.

Irving, E., and Pullaiah, G., 1976, Reversals of the geomagnetic field, magnetostratigraphy, and relative magnitude of palaeosecular variation in the phanerozoic, *Earth Sci. Rev.*, **12**, 36–64.

Irving, E., 1980, Pole positions and continental drift since the Devonian, in M. W. McElhinny (Ed.), *The Earth — its Origin, Structure and Evolution*, Chap. 17, Academic Press, New York.

Isacks, R., Oliver, J., and Sykes, L. H., 1968, Seismology and the new global tectonics, *J. Geophys. Res.*, **73**, 5855–99.

Jenkyns, H. C., 1976, Sediments and sedimentary history of the Manihiki Plateau, South Pacific Ocean, *Init. Rep. DSDP*, **33**, 773–890.

Johnson, D. W., 1931, *Stream Sculpture on the Atlantic Slope*, Columbia, 142 pp.

Johnson, G. L., and Heezen, B. C., 1967, Morphology and evolution of the Norwegian–Greenland Sea, *Deep Sea Res.*, **14**, 755–71.

Karig, D. E., 1970, Ridges and basins of the Tonga–Kermadec island arc system, *J. Geophys. Res.*, **15**, 239.

Karig, D. E., 1971, Origin and development of marginal basins in the western Pacific, *J. Geophys. Res.*, **76**, 2542–61.

Katz, H. R., 1974, Margins of the southwest Pacific, in *The Geology of Continental Margins*, pp. 549–65.

Kelleher, J., and McCann, W., 1976, Buoyant zones, great earthquakes, and unstable boundaries of subduction, *J. Geophys. Res.*, **81**, 4885–96.

Kent, P. E., 1972, Mesozoic history of the east coast of Africa, *Nature*, **238**, 147.

Kent, P. E., 1973, East African evidence of the palaeoposition of Madagascar, in *Implications of Continental Drift in the Earth Sciences*, Academic Press, London, pp. 873–75.

Kent, P. E., 1974, Continental margin of East Africa — a region of vertical movements, in *The Geology of Continental Margins*, pp. 313–20.

Kent, P. E., 1975, Review of North Sea Basin development, *J. Geol. Soc. Lond.*, **131**, 435–68.

Kent, P. E., 1977, The Mesozoic development of aseismic continental margins, *J. Geol. Soc. Lond.*, **134**, 1–18.

King, L. C., 1953, Canons of landscape evolution, *Geol. Soc. Am. Bull.*, **64**, 721–52.

King. L. C., 1956, Rift valleys of Brazil, *Geol. Soc. S. Africa Trans.*, **59**, 199–209.

King, L. C., 1958, Basic palaeogeography of Gondwanaland during the late Palaeozoic and Mesozoic Eras, *J. Geol. Soc. Lond.*, **94**, 47–70.

King. L. C., 1962/7, *The Morphology of the Earth*, 2nd Edn 1967, Oliver and Boyd, Edinburgh.

King, L. C., 1966, Since du Toit: geological relationships between South Africa and Antarctica: 9th Alexander du Toit Memorial Lecture, *Geol. Soc. S. Africa, Spec. Publ.*

King, L. C., 1968, Cymatogeny, in *Encyclopaedia of Geomorphology*, Vol. 3, New York, pp. 240–2.

King, L. C., 1970, Uloa revisited, *Geol. Soc. S. Africa Trans.*, **73**, 151–87.

King, L. C., 1972/82, *The Natal Monocline: Explaining the Origin and Scenery of Natal, S. Africa*, University of Natal Press, Durban, 113 pp.

King, L. C., 1973, An improved reconstruction of Gondwanaland, in *Implications of Continental Drift to the Earth Sciences*, Academic Press, London, pp. 851–63.

King, L. C., 1976, Planation remnants upon high lands, *Z. Geomorph.*, **20**, 133–48.

King, L. C., 1977, The geomorphology of central and southern Africa, in *Biogeography and Ecology of Southern Africa*, pp. 3–17.

Kinsman, D. G. J., 1975, Rift valley basins and the sedimentary history of trailing continental margins, in *Petroleum and Global Tectonics*, Princeton University Press.

Kowsmann, R., Leyden, R., and Francisconi, 1977, Marine seismic investigations, southern Brazil margin, *A.A.P. Geol. Bull.*, **61**, 546–57.

Krause, D. C., 1966, Tectonics, marine geology and bathymetry of the Celebes Sea – Sula Sea region, *Geol. Soc. Am. Bull*, **77**, 813–32.

Larson, R. L., 1976, Late Jurassic and early Cretaceous evolution of the western central Pacific Ocean, *J. Geomag. Geoelectr.*, **28**, 219–36.

Larson, R. L., and Chase, C. G., 1972, Late Mesozoic evolution of the western Pacific Ocean, *Geol. Soc. Am. Bull.*, **83**, 3627–44.

Larson, R. L., and Hilde, T. W. C., 1975, Revised time scale of magnetic reversals for the early Cretaceous and late Jurassic, *J. Geophys. Res.*, **80**, No. 17, 2586–94.

Larson, R. L., and Pitman, W. C., 1972, World wide correlation of Mesozoic magnetic anomalies and its implications, *Geol. Soc. Am. Bull.*, **83**, 3545–662.

Laughton, A. S., Sclater, J. G., and McKenzie, D. P., 1972, The structure and evolution of the Indian Ocean, in *Implications of Continental Drift to the Earth Sciences*, Academic Press, London, pp. 203–12.

Lidmar-Bergstrom, K., 1982, Pre-Quaternary geomorphological evolution in southern Fennscandia, *Med. Fan. Lunds Universiteits Geog. Institute*, **XCI**.

Lisitzin, A. P., 1972, Sedimentation in the World Ocean, *Soc. Econ. Palae. Mineralogists, Spec. Publ. No.* 17, 132 pp.

Lister, L., 1976, The erosion surfaces of Rhodesia, *D. Phil. Thesis, University of Rhodesia.*

Logatchev, N. A., Beloussov, V. V., and Milanovsky, E. E., 1972, East African rift development, *Tectonophys.*, **15**, 71–81.

Long, R. E., and Backhouse, R. W., 1976, The structure on the western flank of the Gregory Rift, Part II. The mantle, *Geophys. J. R. Astron. Soc.*, **44**, 677–88.

Lowell, J. D., Genik, G. J., Nelson, T. H., and Tucker, P. M., 1975, Petroleum and plate tectonics of the southern Red Sea, in *Petroleum and Global Tectonics*, Princeton University Press, pp. 129–256.

Ludwig, W. J., 1970, The Manila Trench and West Luzon Trench, *Deep Sea Res.*, **17**, 553–71.

McBirney, A. R., and Gass, I. G., 1967, Relations of oceanic volcanic rocks to mid-ocean rises and heat flow, *Earth Planet. Sci. Lett.*, **2**, 265–76.

McConnell, R. B., 1971, The association of the East African rift system with Precambrian structures, *Geol. Soc. Lond.*, Circular 166, p. 2.

McConnell, R. B., 1972, Geological development of the rift system of eastern Africa, *Geol. Soc. Am. Bull.*, 83, 2549–72.

McKenzie, D. P., 1970, Plate tectonics and continental drift, *Endeavour*, **29**, 39–44.

McElhinny, M. W., and Brock, A., 1975, A new palaeomagnetic result from East Africa and estimates of the Mesozoic palaeoradius, *Earth Planet. Sci. Rev. Lett.*

McElhinny, M. W., Hailey, N. S., and Crawford, A. R., 1974, Palaeomagnetic evidence shows Malay Peninsula was not a part of Gondwana, *Nature*, **252**, 641–5.

McKenzie, D. P., Davies, D., and Molnar, P., 1970, Plate tectonics of the Red Sea and East Africa, *Nature*, **226**, 243–8.

McKenzie, D. P., and Morgan, W. J., 1969, Evolution of triple junctions, *Nature*, **224**, 125–33.

McKenzie, D. P., and Parker, R. L., 1967, The North Pacific: an example of tectonics on a sphere, *Nature*, **216**, 1276–80.

Maguire, P. K., and Long, R. E., 1976, The structure on the western flank of the Gregory Rift (Kenya), Part 1. The crust, *Geophys. J. R. Astron. Soc.*, **44**, 661–75.

Malin and Saunders, 1974, Rotation of the geomagnetic field, *Nature*, **248**, 403–5.

Mantura, A. J., 1972, Geophysical illusions of continental drift, *A.A.P. Geol. Bull.*, **56**, 1552–6, 2451–5.

Maxwell, A. E., 1969, Recent deep-sea drilling results from the South Atlantic, *Trans. Am. Geophys. Un.*, **50**, No. 4, p. 113.

Maxwell, A. E., Von Herzen, R. P., Hsu, K. J., Andrews, J. E., Saito, T., Percival, S. F., Milow, E. D., and Boyce, R. E., 1970, Deep sea drilling in the South Atlantic, *Science*, **168**, No. 3935, 1047–59.

Maxwell, J. C., 1968, Continental drift and a dynamic earth, *Am. Scientist*, **56**, No. 1, 35–51.

Maxwell, J. C., 1973, The new global tectonics: pro and con, *Geotimes*, **18**, 31.

Melhorn, W. N., and Edgar, D. E., 1976, The case for episodic continental-scale erosion surfaces: a tentative geodynamic model, *Geomorphic Symp.*, Binghamtown, NY, Chap. 13, pp. 243–73.

Menard, H. W., 1961, The East Pacific Rise, *Science*, **132**, 1737–46.

Menard, H. W., 1964, *Marine Geology of the Pacific*, McGraw-Hill, New York, 271 pp.

Menard, H. W., and Atwater, T. M., 1968, Changes in direction of sea floor spreading, *Nature*, **219**, 463–7.

Menke, W. H., and Jacob, K. H., 1976, Seismicity patterns in Pakistan and northwestern India associated with continental collision, *Seismol. Soc. Am. Bull.*, **66**, 1695–711.

Meyerhoff, A. A., and Meyerhoff, H. A., 1972a, The new global tectonics: major inconsistencies, *A.A.P. Geol. Bull.*, **56**, 269–336.

Meyerhoff, A. A., and Meyerhoff, H. A., 1972b, The new global tectonics: age of linear magnetic anomalies of ocean basins, *A.A.P. Geol. Bull.*, **56**, 337–63.

Mitchell, A. H. G., 1973, Composition of olivine, silica activity and oxygen frigacity in Kimberlite, *Lithos*, **6**, 65–81.

Mitchell, A. H. G., 1981, Phanerozoic plate boundaries in mainland S.E. Asia, the Himalayas and Tibet, *J. Geol. Soc. Lond.*, **138**, 109–22.

Mitchell, A. H. G., and Garson, M. S., 1976, Mineralization at plate boundaries, *Min., Sci. & Engng*, **8**, 129–69.

Molnar, P., and Sykes, L. R., 1969, Tectonics of the Caribbean and Middle America Regions, *Geol. Soc. Am. Bull.*, 80, 1639–84.

Mohr, P. A., 1971, Outline tectonics of Ethiopia, *Tectonics of Africa*, UNESCO, Paris, p. 441.

Mohr, P. A., 1973, Crustal Deformation Rate and the Evolution of the Ethiopian Rift, *Implications of Continental Drift to the Earth Sciences*, Academic Press, London, pp. 767–76.

Moore, T. C., 1972, DSDP: successes, failures, proposals, *Geotimes*, 17, No. 7.

Moores, E., 1970, Ultramafics and orogeny with models of the US Cordillera and the Tethys, *Nature, Lond.*, **228**, 837–42.

Morgan, W. J., 1968, Rises, trenches, great faults and crustal blocks, *J. Geophys. Res.*, **73**, No. 6, 1959–82.

Mueller, S., Peterschmidt, E., Fuchs, K., and Amsorge, J., 1969, Crustal structure beneath the Rhine graben, from seismic refraction and reflection measurements, *Tectonophys.*, **8**, 529–42.

Musset, A. E., Brown, G. C., Eckford, M., and Charlton, S. R., 1972, The British Tertiary igneous province: K–Ar ages of some dykes and lavas from Mull, Scotland, *Geophys. J. R. Astron. Soc.*, **30**, 405–14.

Nafe, J. E., and Drake, C. L., 1969, North Atlantic — geology and continental drift, *A.A.P. Geol. Mem.* 12, 59–87.

Norton, I. O., and Sclater, J. G., 1979, A model for the evolution of the Indian Ocean and the break-up of Gondwanaland, *J. Geophys. Res.*, **84**, 6803–30.

Norvick, M. S., 1979, The tectonic history of the Banda Arcs, eastern Indonesia, *J. Geol. Soc., Lond.*, **136**, 519–27.

Oliver, J., Isacks, B. L., Barazangi, M., 1974, Seismicity at continental margins, in *The Geology of Continental Margins*, Berlin.

Ollier, C., 1978, *Tectonics and Geomorphology of the Eastern Highlands in Landform Evolution in Australasia*, Australian National University Press.

Opdyke, N. D., Kent, D. V., and Lowrie, W., 1973, Details of magnetic polarity transitions recorded in a high deposition rate deep-sea core, *Earth Planet. Sci. Lett.*, **20**, 315–24.

Osmaston, M. F., 1973, Limited lithosphere separation as a main cause of continental basins, continental growth and epeirogeny in *Implications of Continental Drift to the Earth Sciences*, Academic Press, London.

Ostenso, N. A., 1973, Sea-floor spreading and the origin of the Arctic Ocean Basin, in *Implications of Continental Drift in the Earth Sciences*, Academic Press, London, pp. 165–73.

226

Pallister, 1971, in *Tectonics of Africa*, UNESCO, Paris.
Perfit, M. R., 1976, Petrology and geochemistry of mafic rocks from the Cayman Trench, *Geology*, **5**, 105–10.
Picard, L., 1970, On Afro–Arabian tectonics, *Geol. Rund.*, **59**, 337–81.
le Pichon, X., 1968, Sea-floor spreading and continental drift, *J. Geophys. Res.*, **73**, 3661–97.
le Pichon, X., and Heirtzler, J. R., 1968, Magnetic anomalies in the Indian Ocean and sea floor spreading, *J. Geophys. Res.*, **73**, 2101–18.
le Pichon, X., Sibuet, J. C., and Francheteau, J., 1977, The fit of the continents around the North Atlantic Ocean, *Tectonophys.*, **38**, 169–210.
Pierce, K., 1965, Geomorphic significance of a Cretaceous deposit in the Great Valley off southern Pennsylvania, *US Geol. Surv. Prof. Paper* 525–C, 152–6.
Pitcher, W. S., 1975, Steady plate motion and episodic orogeny and magmatism, *J. Geol. Soc. Lond.*, **131**, 590.
Pitman, W. C., 1978, Relationship between eustacy and stratigraphic sequences of passive margins, *Geol. Soc. Am. Bull.*, **89**, 1389–403.
Pitman, W. C., Herron, E. M., and Heirtzler, J. R., 1968, Magnetic anomalies in the Pacific and sea floor spreading, *J. Geophys. Res.*, **73**, No. 6, 2069–85.
Pullaiah, G., Irving, E., Buchan, K. L., and Dunlop, D. J., 1975, Magnetization changes caused by burial and uplift, *Earth Planet. Sci. Lett.*, **28**, 133–43.
Press, F., 1969, The sub-oceanic mantle, *Science*, **165**, 174–6.
Quennell, A. M., 1958, The structural and geomorphic evolution of the Dead Sea Rift, *Q.J. Geol. Soc. Lond.*, **114**, 11–24.
Rabinowitz, P. D., and la Brecque, J. L., 1979, The Mesozoic South Atlantic Ocean and evolution of its continental margins, *J. Geophys. Res.*, **84**, 5973–6002.
Rezonor, I. A., 1978, Major sinkings of the ocean floor and the constancy of sea-level, *Int. Geol. Rev.*, **21**, No. 5, 509–16.
Richter, F. M., 1977, On the driving mechanism of plate tectonics, *Tectonophys.*, **38**, 61–88.
Ringwood, A. E., 1974, The petrological evolution of island arc systems, *J. Geol. Soc. Lond.*, **130**, 183–204.
Rona, P. A., 1970, Comparison of continental margins of eastern North America at Cape Hatteras and northwestern Africa at Cap Blanc, *A.A.P. Geol. Bull.*, **54**, 129–57.
Rutland, R. W. R., 1968, A tectonic study of part of the Philippine fault zone, *Q.J. Geol. Soc. Lond.*, **123**, 293–325.
Rutland, R. W. R., 1971, Andean orogeny and sea floor spreading, *Nature*, **233**, 252–5.
Saggerson, E. P., and Baker, B. H., 1965, Post Jurassic erosion surfaces in eastern Kenya and their deformation in relation to rift structure, *Q.J. Geol. Soc. Lond.*, **121**, 51–72.
Saito, T., Ewing, M., and Burckie, L., 1966, Tertiary sediment from the mid-Atlantic Ridge, *Science*, **151**, 1075–9.
Sbar, M. L., and Sykes, L. R., 1973, Contemporary compressive stress and seismicity in eastern North America: an example of intra-plate tectonics, *Geol. Soc. Am. Bull.*, **84**, 1861–82.
Schafer, C., and Brooke, J., 1970, Cores from the crest of the mid Atlantic Ridge, *Geotimes*, **15**, 14–16.
Scholz, C. H., Barazangi, M., and Sbar, M. L., 1971, Late Cenozoic evolution of the Great Basin, western United States, as an ensialic interarc basin, *Geol. Soc. Am. Bull.*, **82**, 2979–90.
Sclater, J. G., and Dietrick, R., 1971, Elevation of mid-ocean ridges and the basement age of JOIDES deep sea drilling sites, *Geol. Soc. Am. Bull.*, **84**, 1547–54.

Searle, R. C., 1970, Lateral extension in the East African rift valleys, *Nature*, **227**, 267–8.

Shive, P. N., 1970, Deformation and Remanence in Magnetite, *Earth Planet. Sci. Lett.*, **7**, p. 451.

Siesser, W. G., Scrutton, R. A., and Simpson, E. S. W., 1974, Atlantic and Indian Ocean margins of southern Africa, in *The Geology of Continental Margins*, New York.

Smith, A. G., and Hallam, A., 1970, The fit of the southern continents, *Nature*, **225**, 139–44.

Sougy, J., 1962, West African fold belt, *Geol. Soc. Am. Bull.*, **73**, 871–6.

Stevens, G., 1980, *New Zealand Adrift*, A. H. & A. W. Reid, Wellington, NZ.

Stocklin, J., 1965, A review of the structural history and tectonics of Iran, *Geol. Surv. of Iran*, 23.

Stocklin, J., 1974, Possible ancient continental margins in Iran, in *The Geology of Continental Margins*, pp. 873–87.

Stoneley, R., 1974, Evolution of the Continental Margins bounding a former southern Tethys, in *The Geology of Continental Margins*, pp. 889–903.

Stoneley, R., 1981, The geology of the Kuh-e Dalneshin area of southern Iran and its bearing on the evolution of southern Tethys, *J. Geol. Soc. Lond.*, **138**, 509–26.

Sykes, L., 1967, Mechanism of earthquakes and nature of faulting on the mid-ocean ridges, *J. Geophys. Res.*, **72**, 2131–53.

Talwani, M., Windisch, C. C., and Langseth, M. G., 1971, Reykjanes Ridge crest: a detailed geophysical study, *J. Geophys. Res.*, **76**, 473–517.

Talwani, M., and Eldholm, O., 1977, Evolution of the Norwegian–Greenland Sea, *Geol. Soc. Am. Bull.*, **88**, 969–99. *Tectonics of Africa*, 1971, UNESCO, Paris.

du Toit, A. L., 19373 *Our Wandering Continents*, Oliver & Boyd, Edinburgh.

du Toit, S. R., and Leith, M. J., 1974, The J(c)-1 borehole on the continental shelf near Stanger, Natal, *Geol. Soc. S. Africa Trans.*, **77**, 247–54.

Van Bemmelen, (1949), *The Geology of Indonesia*, Government Printing Office, The Hague.

Vine, F. J., 1966, Spreading of the ocean floor: new evidence, *Science*, **154**, 1405–15.

Vine, F. J., and Matthews, D. H., 1963, Magnetic anomalies over ocean ridges, *Nature*, **199**, 947–9.

Vinogradov, A. P., *et al.*, 1969, The structure of the mid-oceanic rift zone of the Indian Ocean and its place in the world rift system, *Tectonophys.*, **8**, 377–401.

Vogt, P. R., Avery, O. E., Schneider, E. D., Anderson, C. N., and Bracey, D. R., 1969, Discontinuities in sea-floor spreading, *Tectonophys.*, **8**, 285–317.

Vogt, P. R., and Sweeney, J. F., 1978, Arctic basin morphology, *Polarfirschung*, **48**, 20–30.

Ward, P. L., 1971, New interpretation of the geology of Iceland, *Geol. Soc. Am. Bull.*, **82**, 2991–3012.

Watkins, N. D., 1969, Crustal spreading: a critical comparison of hypothesis requirements and present knowledge of relevant magnetic properties and polarity history (abstract), *Tectonophys.*, **7**, Nos 5–6, 543.

Watkins, N. D., and Kenneth, J. P., 1973, Response of deep-sea sediments to changes in physical oceanography resulting from separation of Australia and Antarctica, in *Implications of Continental Drift on the Earth Sciences*, Academic Press, London, pp. 787–840.

Weissel, J. K., and Hayes, D. E., 1971, Asymmetric sea-floor spreading south of Australia, *Nature*, **231**, 518–21.

Wellman, H. W., 1973, New Zealand fault zones and sea-floor spreading, in *The Western Pacific: Island Arcs, Marginal Seas. Geochemistry*, pp. 335–48.

Westbrook, G. K., Griffiths, D. H., and Barker, P. F., 1975, Atlantic Island Arcs, in *Geodynamics Today: A Review of the Earth's Dynamic Processes*, Royal Society of London.

Whiteman, J. A., 1971, Structural Geology of Sudan, in *Tectonics of Africa*, UNESCO, Paris, pp. 433–45.

Wilson, J. T., 1965, A new class of faults and their bearing on continental drift, *Nature*, **207**, 343–7.

Wilson, J. T., 1969, Aspects of the different mechanics of ocean floors and continents, *Tectonophys.*, **8**, 281–3.

Windley, B. F., 1979, Oceanic and continental transform faults, *J. Geol. Soc. Lond.*, **136**, 267–8.

Wiseman, J. D. H., and Sewell, R. B., 1937, The floor of the Arabian Sea, *Geol. Mag.*, **74**, 219–30.

de Wit, M. J., 1977, The evolution of the Scotia Arc as a key to the reconstruction of southwestward Gondwanaland, *Tectonophys.*, **37**, 53–81.

Worzel, J. L., 1965, Deep structure of coastal margins and mid-oceanic ridges, *Univ. Bristol, Colston Papers*, **XVII**, 335–61.

Wyllie, P. J., 1977, Effects of H_2O and CO_2 on magma generation in the crust and mantle, *J. Geol. Soc. Lond.*, **134**, 215–34.

Yole, R. W., and Irving, E., 1980, Displacement of Vancouver Island, *Can. J. Earth Sci.*, **17**, 1210–28.

York, D., 1982, Part of British Columbia may be Californian, *Globe & Mail*, Vancouver, 22 February.

Zhivago, A. B., 1971, Geomorphological problems of the Southern Oceans. Geomorphology 11693, *Acad. Nauk* (in Russian).

Zÿderveld, J. D. A., and van der Voo, R., 1973, Palaeomagnetism in the Mediterranean Area, in *Implications of Continental Drift to the Earth Sciences*, Academic Press, London, pp. 133–64.

Addenda to References

added at proof stage.

1. Stocklin, V., 1982, The Geology of Nepal and its regional frame. (Thirty-third William Smith Lecture). *J. Geol. Soc. London*, **139**, pp. 1–34
2. 1982, Thematic set of papers on West Antarctica and the Southern Andes. *J. Geol. Soc. London*, **139**, pt. 6.
3. 1983, Report on the February Symposium, in Sydney, *on The Expanding Earth*. ed. S. Warren Carey, with valuable papers from many authors.

Index